Financing Energy Projects in Developing Countries

Financing Energy Projects in Developing Countries

Hossein Razavi, PhD

PennWell®

Disclaimer

The recommendations, advice, descriptions, and the methods in this book are presented solely for educational purposes. The opinions in this book are those of the authors and are not attributable to CRA International, Inc., or the University of Houston. The author and the publisher assume no liability whatsoever for any loss or damage that results from the use of any of the material in this book. Use of the material in this book is solely at the risk of the reader.

PennWell Corporation
1421 South Sheridan Road
Tulsa, Oklahoma 74112-6600 USA

800.752.9764
+1.918.831.9421
sales@pennwell.com
www.pennwellbooks.com
www.pennwell.com

Director: Mary McGee
Managing Editor: Marla Patterson
Production Manager: Sheila Brock
Production Editor: Tony Quinn
Cover Designer: Charles Thomas
Book Designer: Wes Rowell

Library of Congress Cataloging-in-Publication Data;

Razavi, Hossein.
Financing energy projects in developing countries / Hossein Razavi.
p. cm.
Includes bibliographical references and index.
ISBN 978-1-59370-124-6
1. Energy industries—Developing countries—Finance. 2. Energy development—Developing countries—Finance. 3. Economic development projects—Developing countries—Finance. I. Title.
HD9502.D442R39 2007
333.7909172'4--dc22

2007032121

Printed in the United States of America
1 2 3 4 5 11 10 09 08 07

Contents

Executive Overviews

The summaries

- Provide a quick overview for the first-time reader
- Serve as reference notes for the reader who has studied the text

Figures and tables complement the summaries and can be read independently of the text.

Summaries

Figures

Tables

Preface

This book provides firsthand information and analysis regarding how multilateral, bilateral, and commercial financiers make decisions about oil, gas, and power projects. The financing of energy projects in developing countries is challenging in relation to each country's business environment, government behavior, and political risk. Hence, putting together an attractive financing package requires an intimate understanding of all of the following:

- The availability of many sources of soft loans, credits, grants, tied loans, and untied loans
- The objectives, tendencies, and requirements of various financiers
- The possibilities for combining borrowing and guarantee instruments
- The effective ways of approaching financiers

This book describes the intricacies of public and private financing of energy projects and provides guidance in preparing upstream and downstream oil and gas projects, as well as power generation, transmission, and distribution projects. It is divided into three parts: Part I comprises two brief chapters that provide a background of project and corporate financing and the challenges of funding projects located in developing countries. Part II consists of five chapters that introduce the major sources of funding for oil, gas, and power projects. The sources discussed include global multilateral institutions; regional development banks; bilateral aid, credit, and insurance agencies; and commercial capital markets. Part III consists of four chapters that provide guidelines for the preparation of successful project packages. These guidelines include methods for the analysis of business environments, economic and financial viability, financial structures, and environmental concerns for oil, gas, and power projects.

It has been more than 10 years since the first edition of this book was published. The intervening decade has seen a bust and a boom in energy investment and energy finance. As a result, many new players have entered the business of financing energy projects in emerging economies, and many new methods have emerged. These changes necessitated the creation of a new edition of this book. This edition addresses the new business environment by discussion and case studies throughout—particularly in the sections devoted to the description of financiers and the presentation of bankable projects.

Hossein Razavi
The World Bank
Washington, DC
October 2007

ACKNOWLEDGMENTS

This book is based on what I have learned in working with many people from commercial banks, bilateral financing agencies, multilateral organizations, consulting and engineering firms, and public and private oil, gas, and power companies. I am deeply indebted to these people, who are too numerous to name here. Nevertheless, I would like to thank Clive Armstrong, Hossein Askari, Eric Daffern, Bertrand de Frondeville, Robin Glantz, Luis Gutiérrez, Bjørn Håmsø, Georges Khoury-Haddad, Shigeru Kubota, Farida Mazhar, John O'Reilly, William Porter, Kyoichi Shimazaki, Ramin Shojai, Bent Svensson, Barbara Treat, John Treat, Elizabeth Wang, and Adam Wilson for their comments on various drafts of this manuscript and the previous edition of the book. Responsibility for any errors of fact or judgment is solely mine. Finally, I want to thank Evalyne Tandon and Larisa Marquez for their invaluable assistance in preparing the manuscript.

Abbreviations and Acronyms

ACP: African, Caribbean, and Pacific countries EU program

ADB: Asian Development Bank

ADF: Asian Development Fund

ADR: American depository receipt

AFD: French Development Agency (l'Agence Française de Développement)

AfDB: African Development Bank

AIF: African Infrastructure Fund

ALA: Asian and Latin American countries EU program

AMEX: American Stock Exchange

ASEAN: Association of Southeast Asian Nations

BADEA: Arab Bank for Economic Development in Africa (Banque Arabe pour le Développement Economique en Afrique)

BMCE: Banque Marocaine du Commerce Exterieur

BMZ: Federal Ministry for Economic Cooperation and Development (Bundesministerium für Wirtschaftliche Zusammenarbeit), Germany

BOD_5: Biochemical oxygen demand, 5-day test

BOO: Build-own-operate

BOOT: Build-own-operate-transfer

BOT: Build-operate-transfer

BP: British Petroleum

CARIFA: Caribbean Basin Projects Financing Authority

CDC: Capital for Development, United Kingdom

CDM: Clean Development Mechanism

CFCs: Chlorofluorocarbons

CFU: Carbon Finance Unit, World Bank

CIDA: Canadian International Development Agency

CIRRs: Commercial interest reference rates

CNOOC: Chinese National Oil and Gas Company

CNPC: China National Petroleum Corporation

CO: Carbon monoxide

CO_2: Carbon dioxide

COD: Chemical oxygen demand

COFACE: Compagnie Française d'Assurance pour le Commerce Extérieur, France

DEA: Federal Department of Economic Affairs, Switzerland

DFID: Department for International Development, United Kingdom

DGCS: Directorate General for Development Cooperation, Italy

DGIS: Directorate General for International Cooperation, Netherlands

DREE: Direction des Relations Economiques Extérieures, France

DSCR: Debt-service coverage ratio

EA: Environmental assessment

EBRD: European Bank for Reconstruction and Development

EC: European Community

ECGD: Export Credit Guarantee Department, United Kingdom

EDC: Export Development Corporation, Canada

EDCF: Economic Development Cooperation Fund, South Korea

EDF: European Development Fund

EEC: European Economic Community

EFC: Danish Export Finance Corporation

EIA: Energy Information Administration, U.S. Department of Energy

EIB: European Investment Bank

EIU: Economist Intelligence Unit, London

EKR: Export Credit Council (Eksportkreditraadet), Denmark

EMF: Electromagnetic field

EPC: Engineering, procurement, and construction

ERG: Export Risk Guarantee Agency, Switzerland

ESMAP: Energy Sector Management Assistance Programme

EU: European Union

FDI: Foreign Direct Investment

FEMIP: Facility for Euro-Mediterranean Investment and Partnership

FGD: Flue-gas desulfurization

FSU: Former Soviet Union countries

GCC: Gulf Cooperation Council

GDP: Gross domestic product

GDR: Global depositary receipts

GEF: Global Environment Facility

GNI: Gross national income

GNP: Gross national product

GPW: Gross product worth

GTZ: Technical Assistance Corporation (Gesellschaft für Technische Zusammenarbeit), Germany

H_2S: Hydrogen sulfide

IBRD: International Bank for Reconstruction and Development

ICD: Islamic Corporation for the Development of the Private Sector

ICIEC: Islamic Corporation for the Insurance of Investment and Export Credit

ICSID: International Centre for Settlement of Investment Disputes, World Bank

IDA: International Development Association

IDB: Inter-American Development Bank

IDFC: Infrastructure Development Finance Company

IEA: International Energy Agency

IEF: International Energy Finance

IFC: International Finance Corporation, World Bank Group

IFU: Industrialization Fund for Developing Countries, Denmark

IIC: Inter-American Investment Corporation, IDB

IIF: IDB Infrastructure Fund

IMF: International Monetary Fund

IOC: International oil company

IPO: Initial public offering

IPP: Independent power producer

IRR: Internal rate of return

IsDB: Islamic Development Bank

ITCF: International Islamic Trade Finance Corporation

ITRI: Islamic Research and Training Institute

JBIC: Japan Bank for International Cooperation

JCI: Japan Consulting Institute

JEXIM: Export-Import Bank of Japan

JICA: Japan International Cooperation Agency

JNOC: Japan National Oil Corporation

KEIC: Korea Export Insurance Corporation

KExim: Export-Import Bank of Korea

KFAED: Kuwait Fund for Arab and Economic Development

KfW: Kreditanstalt für Wiederaufbau, Germany

KOICA: Korean International Cooperation Agency

LIBOR: London Interbank Offer Rate

LNG: Liquefied natural gas

LPG: Liquefied petroleum gas

LRMC: Long-run marginal cost

LSTK: Lump-sum turnkey contract

MCA: Millennium Challenge Account, United States

MCC: Millennium Challenge Corporation, United States

MEDA: Euro-Mediterranean Partnership, an EU program

MENA: Middle East and North Africa

MIF: Multilateral Investment Fund, IDB

MIGA: Multilateral Investment Guarantee Association, World Bank Group

MITI: Ministry of International Trade and Industry, Japan

MLA: Mandated lead arranger

MOFA: Ministry of Foreign Affairs, Japan

MSDS: Material safety data sheet

NDF: Nordic Development Fund

NEFCO: Nordic Environmental Finance Corporation

NEPA: National Environmental Policy Act, United States

NEXI: Nippon Export and Investment Insurance

NG: Natural gas

NGL: Natural gas liquids

NGO: Nongovernmental organization

NIB: Nordic Investment Bank

NOC: National oil company

NOPEF: Nordic Project Fund

NORAD: Norwegian Agency for Development Cooperation

NO_x: Nitrogen oxides

NPV: Net present value

NPW: Net product worth

NYSE: New York Stock Exchange

O&M: Operations and maintenance

OAU: Organization of African Unity

OCT: Overseas countries and territories EU program

ODA: Official development assistance

OECD: Organisation for Economic Co-operation and Development

OECF: Overseas Economic Cooperation Fund, Japan

OPEC: Organization of Petroleum Exporting Countries

OPIC: Overseas Private Investment Corporation, United States

PCBs: Polychlorinated biphenyls

PM: Particulate matter

PM_{10}: Particulate matter of 10 microns or less in size

PPIAF: Public-Private Infrastructure Advisory Facility

PSDU: Private Sector Development Unit, AfDB

PV: Present value

R&D: Research and development

S&P: Standard & Poor's

SABIC: Saudi Arabia Basic Industries Corporation

SACE: Sezione Speciale per Assicurazione del Credito all'Esportazione, Italy

SAMIR: Societe Anonyme Marocaine de l'Industrie du Raffinage

SAREC: Swedish Agency for Research Cooperation with Development Countries

SDC: Swiss Agency for Development and Cooperation

SEC: Securities and Exchange Commission, United States

SECO: State Secretariat for Economic Affairs, Switzerland

SFD: Saudi Fund for Development

SFO: Fund for Special Operations, IDB

SIDA: Swedish International Development Authority

SMEs: Small and medium enterprises

SO_2: Sulfur dioxide

SO_x: Sulfur oxides (a mix of SO_2 and other oxides of sulfur, e.g., sulfur pentaoxide [SO_5])

SPA: Sichuan Petroleum Administration

SPV: Special-purpose vehicle

T&D: Transmission and distribution

TACIS: Technical Assistance to the Commonwealth of Independent States, an EU program

TDA: U.S. Trade and Development Agency

TOC: Total organic carbon

UNDP: United Nations Development Programme

USAID: U.S. Agency for International Development

USExim: U.S. Export-Import Bank

Units and Conversion Factors

BCM: Billion cubic meters

b/d: Barrels per day

bl: Barrel

bp: Basis points

BTU: British thermal unit

GWh: Gigawatt-hour

km: Kilometer

kV: Kilovolt

kWh: Kilowatt-hour

MCF: Thousand cubic feet

mg: Milligram

MMBTU: Million British thermal units

MMCFD: Million cubic feet per day

MW: Megawatt

MWe: Megawatt electric

ng: Nanogram

ng/J: Nanogram per joule

Nm^3: Normal cubic meter

ppm: Parts per million

Rmb: Renminbi (Chinese yuan)

t/d: Tons/day (or tonnes/day)

Note: All dollar ($) figures are U.S. dollars unless otherwise indicated.

Oil

1 bl petroleum = 42 U.S. gallons = 35 Imperial gallons

1 long ton = 1.016 metric tons

1 ton crude oil = 7.33 barrels

1 ton crude oil = 40 MMBTU

1 million tons crude oil/year = 20,000 b/d

1 ton fuel oil = 6.7 barrels

Gas

1 cubic meter = 35.3 cubic feet

1,000 MMCFD = 10 BCM/year

1 cubic meter NG = 36,000 BTU

1 cubic foot NG = 1,030 BTU

1,000 MMCFD NG = 7 million tons of LNG/year

1 ton LNG = 15.5 bl LNG = 9.53 bl crude oil

1 ton LNG = 1,380 cubic meters of NG

1 ton LNG = 52 MMBTU

Electricity

1 MW = 10^6 watts

1 kWh = 3,412 BTU

1 GWh = 1,000,000 kWh

1 BTU = 252 calories = 1,055 joules

1 joule = 0.239 calories

1 petajoule = 10^{15} joules

1 thermie = 10^6 calories = 3,968.3 BTU

1 therm = 100,000 BTU

1 GWh of electricity

- Uses approximately 250 tons of oil in an oil-fired conventional steam power plant
- Uses approximately 390–400 tons of coal in a coal-fired steam power plant
- Uses approximately 8,000 MCF of NG in a combined-cycle power plant

1 mg = 10^{-3} g

1 ng = 10^{-9} g

PART I
THE SETTING

PART I: The Setting

Part I consists of two brief chapters on the basics of financing energy projects in developing countries. It describes

- The concepts of public versus private financing and corporate-based (also known as recourse) versus project-based (nonrecourse) borrowing, as well as general sources of equity and debt financing
- The issues particular to the funding of projects located in developing countries, the implied project risks, and the objectives of risk mitigation

1 Fundamentals Of Project Financing

Before the 1970s, most petroleum projects in developing countries were financed by the internal cash generation of international oil companies (IOCs). During the 1970s, governments became heavily involved in the petroleum sector to ensure better control of their reserves and, in the case of petroleum-importing countries, to quell concerns regarding the security of oil supply. Consequently, oil and gas projects received increased financing from governments' budgets and official borrowings, as well as from IOCs.

Since the early 1990s, most governments have limited their interventions in the petroleum sector and their budgetary contributions thereto in order to encourage the private sector to undertake the required investments. IOCs have again become the main source of investment but often undertake projects with a wide range of partners. As a result, funding of oil and gas projects has become quite complex, involving public and private investors and financiers.

Electricity supply in most developing countries has traditionally been undertaken by a vertically integrated public monopoly that generates, transmits, and distributes power to customer groups in various locations. Electricity supply is often viewed as a socioeconomic service and frequently involves general subsidies by the government, cross-subsidies among various classes of customers, or both. Required investments are generally funded through the utilities' internal resources and official borrowing from multilateral and bilateral agencies. In addition, in many instances, utilities' funds are supplemented by government budgets.

The worldwide move toward further reliance on the private sector is changing the economic structure of the power sector in many developing countries. In some countries, power sector assets are being privatized. In others, power utilities remain government corporations, but the private sector has been invited to build new power projects. Often, the private sector builds a power generating plant and sells its output to the public utility, which retains control of power transmission and distribution (T&D).

T&D networks are still viewed as natural monopolies for which government ownership or regulation is justified. Therefore, some countries have opted to maintain T&D facilities under public ownership, whereas others have aimed, at a minimum, to franchise distribution systems to private investors. For these reasons, new investments in power T&D are likely to be undertaken by both public and private corporations.

Summary 1–1. Public and Private Financing of Energy Projects

Energy projects, particularly gas and power facilities, will likely be subject to both public and private ownership in the foreseeable future. Accordingly, this book covers the sources and requirements for public (government or government-guaranteed) financing and private funding for oil, gas, and power projects.

Recourse and Nonrecourse Financing

In many cases, energy sector investments are funded by corporations (public or private) on the basis of credibility, income, and assets of the corporation. A typical example would be investment in the construction of a new power plant. The funds for such an investment would come from the internal cash (profit) of the power corporation and borrowing by the corporation. The new project would be built as an extension of the assets of an existing company. Capital investment and borrowing would not be on the project account. Rather, the loans would be considered to be the company's debt, and the lenders would have full recourse to all assets and revenues of the company, not just those related to the new power plant.

The aforementioned arrangement is referred to as *corporate financing*. That is, borrowing is carried out by the company as a whole, rather than by an entity created specifically to hold ownership of the new power plant. Although funds may be provided for investment in a specific project, lenders in a corporate-financing arrangement look to the cash flow and assets of the entire company to service the debt and to provide security. Lenders thus carefully examine the company's financial track record, the viability of its expansion plans, and its projected financial performance.

An alternative to corporate financing is the formation of a project company specifically to construct the power plant. In this case, investments in construction of the power plant are viewed as assets of the project company. These funds come in the form of equity or debt. If the company is privately held, the equity is provided by the private owners. Additional funds are borrowed from a variety of sources. In such arrangements, the project's assets and cash flow secure the debt, and creditors do not have recourse to the sponsors' other available resources. Thus, borrowing for the project is not recorded against a sponsor's general balance sheet or creditworthiness. Such self-standing borrowing, with no guarantees from sponsors (or governments) to lenders, is known as *nonrecourse financing* (it is also called *off-balance-sheet financing* or *project financing*).

In practice, however, most projects have *limited-recourse financing,* for which sponsors commit to providing contingent financial support (above their up-front equity commitment) to give lenders extra comfort. The extra security usually focuses on the construction and startup periods, which are the riskiest times for energy projects. For example, sponsors may agree to repay loans from sources other than the project company's assets (if project-company funds fall short) in cases of delays in project completion or cost overruns. Until the project passes final completion tests, lenders will have recourse to the sponsors' other assets and revenues. After completion, however, the lenders' right to repayment is limited to the income stream from the project and its underlying assets, including plant contracts and permits.

Summary 1–2. The Concepts of Recourse and Nonrecourse Financing

- Gas and power projects in developing countries are traditionally built as extensions of the assets of an existing company. For example, when a public utility constructs a power plant, funds are provided on the account of the entire company rather than on the account of the new plant. As loan security, lenders have full recourse to the assets and revenues of the entire company, rather than recourse only to those funds related to the new power plant.

- An alternative is the formation of a project company specifically for the construction of the power plant. The project sponsors contribute equity. The assets and cash flow of the project itself secure debt, not the sponsors' other available resources. This type of borrowing, with no guarantees by sponsors (or governments), is described as nonrecourse. Since the repayment of the loan is primarily dependent on the success of the project, lenders pay close attention to project risks.

Mobilizing Equity and Debt Finance

Structuring a project-finance package entails deciding how much of the project resources should come in the form of equity; the rest will be project debt. *Equity* refers to funds put into the project company by shareholders of the company. Equity holders are owners of the company, and they receive dividends and capital gains based on net profits. Equity holders take risk; they would receive no dividend payments if the company lost money.

Project debt refers to funds lent to the project company by financiers such as commercial banks, insurance and pension funds, and multilateral institutions. These loans are secured by the project's underlying assets. Lenders receive payments for principal and interest on these loans whether the company makes or loses money. However, prospective lenders examine the company's projected cash flow very carefully to ensure that there is sufficient financial capacity for debt repayment.

A special type of funding called *quasi-equity* is occasionally used to attract risk-averse investors. In this case, a special class of company shares—known as *preferred shares*—is introduced. Payment to holders of preferred shares takes priority over dividends to ordinary shareholders but is subordinated to the claims of debt financiers and other creditors of the company.

Summary 1–3. Equity and Debt Financing

Funds to construct a new plant come in the form of either equity or debt:

- *Equity* is provided by owners, also called *shareholders*. Owners include project sponsors, who are the driving force behind a project, as well as "passive" investors, who hold equity positions without being involved in project promotion. Owners' return on their equity investment depends on the company's profit.
- *Debt* refers to funds lent by financiers. Lenders receive payments according to a predetermined rate (or formula). Their payment does not depend on the profit of the company.

For energy projects, equity varies between 20% and 40% of the project cost. Clearly, a higher equity ratio means a higher commitment by project sponsors and a lower risk for lenders. Thus, lenders like to see high equity ratios, whereas sponsors prefer lower ratios to minimize the funds they lock into one project. The acceptable equity ratio depends on the creditworthiness of the sponsors, the risks, and the location of the project.

Mobilizing equity and debt finance have now become intertwined. To raise the money to finance investments, the project promotes issues or sells securities. These securities (also known as *financial instruments*) are in the form of a claim on the future cash flow of the project and a contingent claim on the assets of the project. Traditionally, a security could be in the form of either debt or equity. Today, there are many innovative financial instruments, some of which share the characteristics of both debt and equity. This latter group, which itself comes in a variety of forms, is called *mezzanine finance*. Besides debt, equity, and mezzanine finance, a project may use other instruments, such as leasing, which is applicable to some aspects of the energy sector (e.g., upstream oil and gas) and practical in certain countries (e.g., when Islamic law prohibits debt issues).

In practice, the type of security is distinguished by the order or priority in claim of the debts over the cash flow and assets of a company. Senior debts rank highest among the financial instruments. These debts are normally in the form of loans from commercial banks, merchant banks, investment banks, development agencies, pension funds, and export credit agencies. The terms and conditions of loans vary from lender to lender and from borrower to borrower. Loans can have fixed or variable (floating) interest rates. Repayment of a loan could be from 7–10 years, for an oil sector project, to 15–20 years, for a power project. The cost of raising debt capital includes certain fees besides the interest:

- Management fee, a percentage of the loan paid up front for managing the debt issue
- Commitment fee, charged on the undrawn portion of the loan
- Underwriting fee, paid up front as a percentage of the loan to the bank or financial institution that guarantees to contribute to the loan if it is not fully subscribed
- Success fee, paid up front as a percentage of the loan
- Guarantee fee, paid annually on the outstanding loan amount if it is guaranteed against default

All or some of these charges may be submerged in the interest rate.

Next to senior debt is subordinated debt, which has second claim to the assets of the project. It often comes from sources that have direct interest in the project or specialized funds and that are comfortable taking more risk than a senior lender would. As expected, subordinated debt can be significantly more expensive than senior debt. The interest rate on a subordinated loan is higher than the senior debt; the premium would depend on various risk parameters.

Finally, equity instruments involve the highest risk. Again, it is often the case that the project sponsors bear the highest risk and some other shareholders may be entitled to a priority claim over the company's assets. The equity can be contributed by project sponsors; by investment funds; by multilateral institutions, such as the International Finance Corporation (IFC); by foreign and domestic institutional investors, such as pension funds; and by issuing shares of the company on the domestic or foreign equity markets.

Mobilizing equity becomes even more complex when a project is a joint venture between private and public corporations. In such a case, the equity share of the public corporation can come from the government and a host of sources of official lending agencies (the World Bank, regional banks, export-import banks, bilateral trade and aid agencies, etc.).

Summary 1–4. Sources of Equity and Debt Funds

Equity makes up 20%–40% of project cost and is provided through

- Sponsors' own capital and subordinated loans
- Multilateral institutions
- International equity markets
- Local capital markets
- Certain investment funds
- Governments, a host of official lending and aid agencies, if the project is a state-owned or joint venture between the private and public sectors

Debt makes up 60%–80% of the project cost and is provided through

- Institutional investors (pension funds, insurance companies, and mutual funds)
- International commercial banks
- The International Finance Corporation (IFC) and regional development banks
- International bond markets
- Local banks and bond markets
- Suppliers' credit
- Specialized energy funds
- Government-guaranteed official loans from multilateral institutions, regional banks, and bilateral agencies, if the project is owned by a state agency or is a joint venture by the private and public sectors

External Sources of Funds

The composition of funds available for investment finance has changed dramatically since the 1990s. Prior to the 1990s, there was a heavy reliance on external finance, particularly funds from multilateral agencies such as the World Bank. In the period between 1990 and 1997, the dependence on external finance increased further, while there was a shift to flows from private sources such as Western investors and international commercial banks. Since 1997, there has been a significant shift to domestic capital as a number of countries have been able to develop their bond and equity markets. These changes are explained in chapter 7, along with relevant examples.

Financial flows from external sources to developing countries (fig. 1–1) are categorized as official and private. Official flows comprise development assistance and borrowing from multilateral and bilateral agencies under government sponsorship and guarantee. Private flows are divided into debt and equity: debt includes bonds and bank borrowing; equity includes foreign direct investment (FDI) and portfolio equity. Until the 1990s, official flows constituted the major source of external funding to developing countries and provided about $50–60 billion per year. The overall level of official flows has remained stable over the past two decades, but the net inflow has fluctuated owing to reverse flows (repayment of loans) to the extent that the net official flow turned negative in 2005.

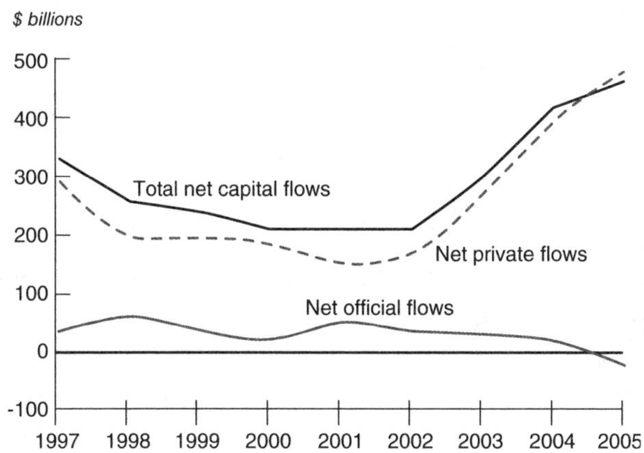

Fig. 1–1. Financial flows to developing countries

The private capital flows, which were at relatively small levels until the 1990s, experienced a rapid growth between 1990 and 1997, when the net inflow reached about $300 billion. In the aftermath of the East Asian and Russian crises of the late 1990s and the collapse of the U.S. equity markets in 2000, the private flows dropped to about $150 billion. Since 2003, a new surge has occurred, increasing the net private capital flows to about $600 billion in 2006. The debt component of private flows was approximately $250 billion and consisted of 42% bonds and 58% bank borrowing. The equity component was about $350 billion, which consisted of 77% FDI and 23% portfolio investments.

During the depressed 1998–2001 period, the FDI component of equity flows remained relatively stable. The portfolio investment and debt flows, by contrast, dropped rapidly—and even turned negative in 2001—because of lenders' attempts to withdraw funds from developing countries. This was most pronounced in bank lending that turned substantially negative in 2001 and 2002. The negative flow affected Latin America and East Asia more than other regions.

The surge of private debt to developing countries during 2003–2006 has surprised many analysts who had concluded that international commercial banks would not be willing to expose themselves to developing countries for many years to come. Indeed, bank lending has grown the fastest during this period. The rapid growth in the volume of private debt flows coincides with a rapidly declining return on these flows. Many of the loan syndications are now priced at a spread of less than 100 basis points (bp) over the London Interbank Offer Rate (LIBOR).

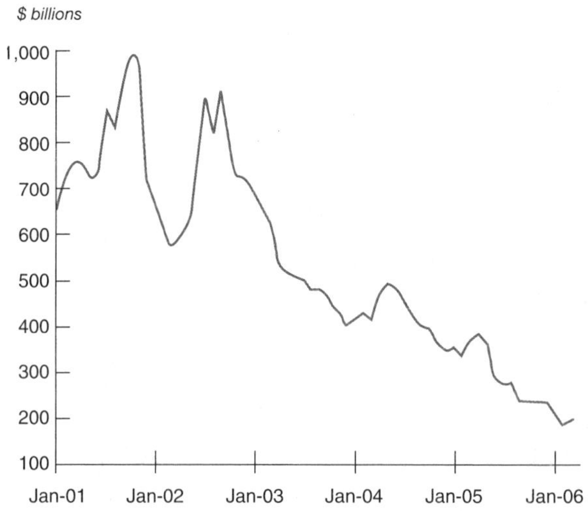

Fig. 1–2. Benchmark spreads (over U.S. treasuries) for emerging market bonds

The bond spreads have also dropped in an unexpected manner (fig. 1–2). Emerging market bond spreads were about 1,200 bp in 1998, in the aftermath of the East Asia crisis and financial turmoil in Latin America. They have consistently dropped since 2001, to about 200 bp in 2006. This sharp decline persisted despite the financial crises in Argentina and Turkey during this period.

The availability of large private finance (equity and debt) at low rates to developing countries is due partly to the apparent abundance of liquidity in the corporate world and partly to the reduced riskiness of lending or investing money in developing countries. Many developing countries have accumulated and hold sufficient reserves of foreign exchange to respond quickly to any signs of financial instability and capital flights.

Project Risks

Essential to structuring a project finance package are identification, analysis, mitigation, and allocation of project risks. These risks are related to events that could endanger the project during development, construction, and operation.

At the stage of project development, the main risk is rejection by the government or by the financiers. Reasons for project rejection include commercial weakness and failure to obtain licenses, permission, and clearance. At the development stage, risks are high but involve relatively small losses, which are limited to money and time spent on feasibility studies and related preparatory work. These risks are borne primarily by project sponsors. However, in many energy projects, sponsors may be able to find multilateral and bilateral grants or other types of aid to finance part of the cost of project development. The idea of mobilizing grants for project preparation/formulation may be instrumental when sponsors cannot afford to risk their own money.

During construction, the main risk is failure to complete the project with acceptable performance levels and within an acceptable time frame and budget. This risk falls mainly on the project company and its sponsors. They in turn hedge their risk by purchasing various forms of insurance and obtaining guarantees from contractors with regard to costs, completion schedule, and operational performance of the project. The construction risk is high and potentially involves significant losses. This is a risk of great importance from the point of view of financiers. Should a project fail during construction, the lender's loan security—that is, the assets of the project company—would be of little value. Thus, financiers do not want to take any of the construction risk and normally ask for recourse to the sponsors' other resources until the project is completed and tested.

After a plant has been constructed, the main concern is that it may not operate on a continuing basis within acceptable economic and technical parameters. Such operational risks are numerous but are usually modest in magnitude. They are related to technical failures, unavailability of fuels, market demand and prices, fiscal issues (taxes or subsidies), foreign exchange rates and convertibility, environmental problems, and so on. These operating risks are borne by the project company. However, a project company can hedge against risks through contractual and guarantee arrangements that in effect transfer some of the risks to other parties. The following are examples of such hedging:

- The project company receives guarantees from equipment suppliers for equipment performance.
- The project company obtains a supply guarantee from a fuel source, at defined prices that might be passed through or, in case of margin shortfall, with netback pricing down to a certain floor.
- The project company receives a take-or-pay contract from a company that buys project output at defined prices, sometimes including a pass-through of certain operating costs (e.g., fuel costs).
- The project company receives guarantees against political risk from multilateral or bilateral agencies, channels foreign revenues through an offshore disbursement account, or both.
- The project company receives limited support from shareholders for defined margin shortfalls, through methods such as cash injection, subordinated loans, or dividend clawbacks.

Hedging arrangements affect not only the liability of the project company but also the willingness of private investors and financiers to support the project. This is the primary reason that project financing has turned into a complex discipline involving a combination of instruments for guarantees, borrowing, and mobilization of equity.

Summary 1–5. Project Risks

- Identification, analysis, mitigation, and allocation of project risks are essential when structuring a project finance package.
- Project risks endanger the project during project preparation, plant construction, and plant operation.
- All risks are initially borne by project sponsors. However, project sponsors enter into numerous guarantee and contractual arrangements to hedge against risks.
- The practice of combining various instruments for guarantees, borrowing, and mobilization of equity has turned into a complex discipline requiring innovative methods of finance.

2 Challenges Of Project Financing In Developing Countries

Structuring financing packages can be especially difficult for projects in developing countries. The main reasons for this are as follows:

- Many developing countries have limited domestic resources. The main shortage is, of course, in capital funds. Other resources—such as skilled labor, raw materials, and infrastructure—may also be scarce.
- Because of the aforementioned shortages, prospective investors need to rely on foreign capital and other resources not locally available.
- To acquire foreign resources, investors must be convinced—and able to convince others (particularly financiers)—that these resources will be utilized safely, efficiently, and profitably and that returns on investments can be repatriated in accordance with known rules and regulations.
- The aforementioned concerns are directly linked to the host country's business environment, including political and economic stability, legislative and regulatory systems, and labor disciplines.
- Imperfections in the country's business environment increase basic project risks and are magnified by a general reluctance to lend to the developing world.

Financing a project in the developing world is considered more risky than financing one in the developed world. Investors want higher returns to compensate for the higher risk. Investors also prefer to share risk by involving others in a project and to take measures to avoid or mitigate project risks. As a result, structuring a finance package involves many parties and instruments. This complexity is certain to increase with time. (See fig. 2–1.)

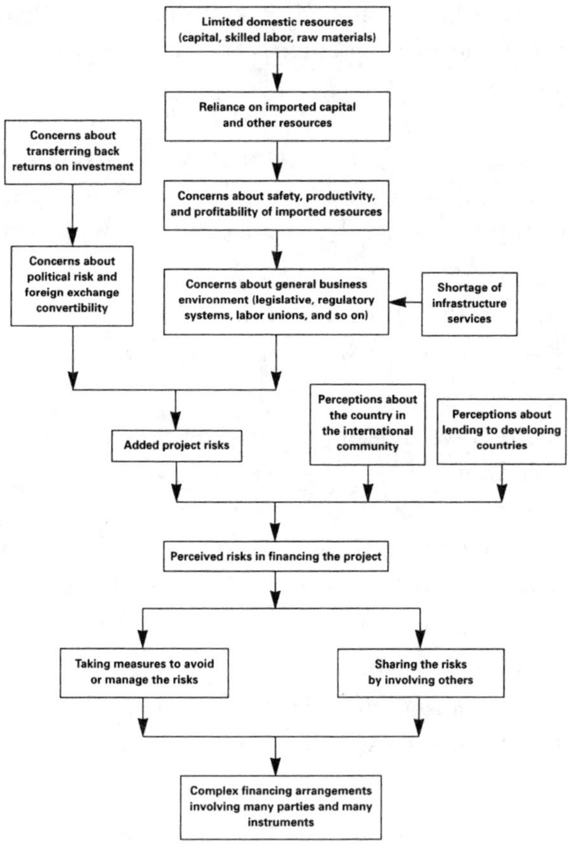

Fig. 2–1. Complexities of financing a project in the developing world

The Differences in Business Environments

Business environments in developing countries differ from those of developed countries mainly in the following areas:

- Institutional and organizational structures and management
- Legislative and regulatory systems
- Economic and political security

Institutional and organizational deficiencies in the energy sector normally relate to structural rigidities or vague divisions of responsibility. For example, the power industry in many developing countries has an integrated structure with a single national utility responsible for generation, transmission, and distribution. This structure is inherently inconsistent with a competitive environment. Although many of these countries try to create conditions that encourage private parties to enter the power sector, rigidities of the integrated system limit the potential for private participation. When private parties do formulate projects, they often find business arrangements difficult to conclude.

Organizational vagueness can delay projects and push costs above budgeted levels. Responsibilities of government agencies and national oil companies (NOCs)—with regard to petroleum development strategy, for example—often overlap. This divided responsibility results in delays in awarding of upstream contracts, construction of gas infrastructure systems, and so on.

Organizational deficiencies often originate in legislative and regulatory systems. Furthermore, these systems are responsible for

- The lack of a transparent and open structure for dealing with private sector activities and, particularly, foreign investors
- Inefficient systems of taxes and duties
- Insufficient and inflexible power tariffs and oil and gas prices
- Unclear safety and environmental standards

Finally, economic and political insecurity are probably the most important reasons for concern about the business environment in developing countries. Economic insecurity affects demand and consumers' ability to pay for project output. Political insecurity can raise costs, reduce revenues, and, in certain cases, result in the total loss of investments or returns on investment. Risks resulting from political insecurity are as follows:

- *Expropriation*. This is the risk that property rights will be seized from the project by the host government, which often occurs without any compensation. Although fear of outright expropriation is often present, investors' greatest fear is creeping expropriation—that is, growth of the host government's take, through increased taxes, royalties, fees, and so on, during the project construction or operation.
- *Breach of undertakings by the host government*. This is the risk that the host government will not fulfill promises stated in the project agreement with regard to contributions. These contributions can be financial ones, or they can be

promises to build necessary infrastructure, such as roads, pipelines, or export terminals. The undertakings may also include such details as labor agreements and removal of certain regulations or governmental fees. Most host governments want to live up to their agreements because their reputations will be tarnished in the global investment community if they do not. However, changes in a country's political or economic landscape sometimes make it impossible to live up to commitments.

- *Civil unrest.* This is the risk that civil disturbances will cause disruption of the project construction or operation. Disruptions can also physically damage the project. Obviously, the consequences to a project's financial return can be disastrous.
- *War.* The risk here extends to the total loss of lenders' and sponsors' capital. Even if complete loss does not occur, export lines could be disrupted, which would diminish the financial attractiveness of the project.
- *Expatriation of profits.* This is the risk that the host country will not allow profits realized from a project to be repatriated to the sponsors' home country. For example, the host country might insist that profits should be invested in other similar projects that may or may not be of financial benefit to the current project sponsors.
- *Inconvertibility of a developing country's currency.* This risk can have a substantial impact on a project that sells much of its product locally.

MITIGATING RISKS

International financiers, particularly commercial ones, have become quite cautious about lending funds to projects in developing countries. The debt crisis of the 1990s is still remembered by the banking and financial communities. Indeed, the banking system in the industrialized world has put strict regulations into effect for lending to developing countries. For example, in the United States, banks must adhere to larger reserve requirements when lending to countries that do not belong to the Organisation for Economic Co-operation and Development (OECD) than they do when lending to member countries; they are not allowed to take the front-end fees that they usually earn as income at the time of the loan, as they can on OECD loans; and they must adjust their portfolios each year as part of their regulatory examination.

Summary 2–1. Major Concerns about Investing in the Developing World

- Financing a project in the developing world is considered more risky than in the developed world because of (1) deficiencies in institutional and organizational structures, (2) lack of clear and transparent legislative and regulatory systems, and (3) economic and political insecurity.
- These risks endanger the viability and sustainability of the project through (1) excessive construction and operation costs, (2) shortfall in revenue or the margin caused by price and market risks, and (3) uncertainty about safety and transferability of investments and returns.

Because of the aforementioned concerns and restrictions, banks seek a complete security package before they finance projects in developing countries. Ideally, the security package should protect the project against all significant risks. These risks are classified in a variety of ways but essentially relate to

- Costs (including construction and operations costs), which may be affected by inflation, interest-rate fluctuations, changing availability and rates of foreign exchange, delays, cost overruns, lack of raw materials or fuels, and so on
- Revenues, which may be affected by price risk and demand risk and which may combine with the costs into margin risk
- Safety and transferability of investments and returns

Accordingly, measures to manage the risks try to convince financiers that

- Costs will not exceed the projected levels, and, if they do, some other party will take the burden before the cost increase affects the financing of the project.
- Revenues will not fall short of projected levels, and, if they do, some other party will make up the shortfall so that project finances are not hurt.
- Investment is safe and returns can be transferred out of the country, or, if funds cannot be transferred, a credible agency will cover legitimate losses.

PART II
GETTING TO KNOW THE FINANCIERS

Before attempting to structure the financing of a project or even approaching financiers, project developers should become familiar with

- The many sources of funds, including grants, soft loans, and technical assistance, that can be tapped for a project.
- The role that each financier normally wishes to play in a project. Most financiers would like to enter a project when the business environment is ready. Others, such as multilateral institutions, want to come in only if they can improve certain aspects of the business environment.

- The orientation of each financier. Some financiers like to support private sector participation, others prefer to promote certain technologies, others yet want to assist poverty reduction, and so on. Familiarity with these tendencies is required not only for choosing the right financier or financiers but also for packaging the project appropriately.
- The methods for combining the facilities provided by various financiers. Often this is necessary for energy projects because these projects are large and because many financiers prefer to join a project only when others are also participating.

Part II of this book familiarizes the reader with financiers and gives instruction in how to deal with them; it consists of five chapters:

- "Sources of Financing" lists possibilities for financing oil, gas, and power projects.
- "Accessing Support from Global Multilateral Institutions" explains the availability of assistance from agencies such as the World Bank and the IFC.
- "Borrowing from Regional Development Banks" describes the roles of the Inter-American Development Bank, the Asian Development Bank, and others.
- "Bilateral Sources of Financing" covers the activities of development assistance programs and export credit agencies.
- "Accessing Commercial Funds" explains the possibilities for mobilizing equity and debt financing from international and domestic capital markets. This chapter also covers various types of guarantees and agreements that can be used to mitigate project risks.

3 Sources Of Financing

Methods and sources for financing energy projects in developing countries have changed significantly over the past two decades. First, commercial banks have become more cautious about lending to developing countries after the debt crises of the 1980s and 1990s. Second, a vast number of new instruments have been introduced to channel financial resources to equity and debt financing for energy investments. Third, energy corporations have developed a tendency not to keep project debt on their balance sheets for long periods. Fourth, governments increasingly prefer not to sponsor borrowing for oil, gas, and power projects.

As a result of such developments, methods of financing have become more diverse (fig. 3–1). These changes have taken a different course in oil projects as compared to power and gas projects.

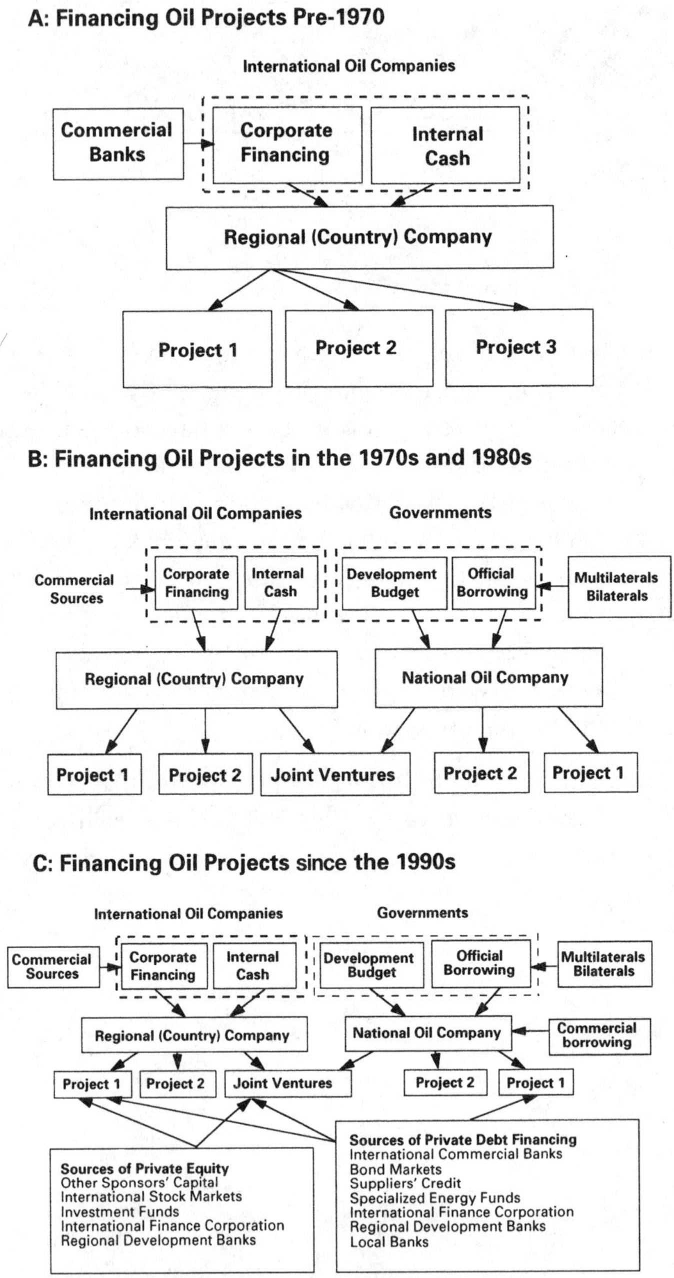

Fig. 3–1. Changes in methods of financing oil projects

Financing Oil Projects

The financing of oil projects in developing countries has evolved over three distinct eras. Before the 1970s, most oil projects were financed by IOCs. Most IOCs had low debt-to-equity ratios; hence, they generally drew funds for exploration and development from internal cash flow and needed to do little borrowing. In addition, if the IOCs did have to borrow, they could raise long-term finance at favorable rates based on their creditworthiness and strong balance sheets.

During the pre-1970s era, the investment behavior of IOCs was determined more by strategic agendas than by strict financial criteria. In the 1970s and the early 1980s, IOCs continued their active role in financing oil investments while governments, in both oil-exporting and oil-importing countries, began taking prominent roles in the petroleum sector. In oil-exporting countries, government involvement was motivated by the incentive to preserve their national resources and ensure the security of their share of oil reserves. In oil-importing countries, interest stemmed from concern that security of oil supply was essential for national economic stability; thus, governments attempted to explore and develop domestic resources and to supervise petroleum imports firmly. Government intervention, enforced mainly through strengthened NOCs, was accompanied by flows of state funds into petroleum projects or state-sponsored borrowing into equity and debt financing of such projects. In this period, IOCs, governments, multilaterals, bilaterals, and commercial banks were quite willing to invest in or finance oil projects.

During the second half of the 1980s and the early 1990s, however, almost all of the aforementioned sources of funds limited their exposure to oil investments. The most important reason was the collapse of oil prices in 1986 and the continuing softness of prices in the 1990s. The lower and uncertain petroleum prices not only reduced the attractiveness of oil investments but also imposed financial hardship on most IOCs. This hardship was exacerbated for some IOCs by new environmental standards for existing facilities, which translated into significant costs. As a result, IOCs restructured and adopted new, stringent investment standards. They increasingly preferred shared risks and faster paybacks than they required in the earlier era. At the same time, the governments of both oil-exporting and oil-importing countries substantially reduced their interventions and support of petroleum investments, to limit the government role in activities that could be carried out by the private sector.

Today IOCs and NOCs are in more comfortable financial positions thanks to the increase in oil prices during the period from 2000 to 2006 but are not willing to go back to the traditional culture of investment finance. Although IOCs and governments still sponsor or provide financial resources for oil projects, a variety of new instruments are regularly employed to fund these projects. Private sources of equity and debt financing increasingly fund oil projects on nonrecourse and limited-recourse bases. Some major oil companies prefer to fund projects from their own resources—including borrowing on their corporate accounts—rather than to use project-based financing. For these companies, with their strong credit ratings, corporate debt is cheaper than project-based borrowing. However, many smaller IOCs are now involved in the petroleum sectors of developing countries, and these companies may not have very strong balance sheets and may not wish to borrow on their corporate accounts. In fact, even major oil companies borrow on a project basis when dealing with very large projects, when facing country-limit considerations (i.e., when a company's exposure in one country is too high or is higher than the company's strategic allowance), or when they are involved with partners that cannot afford to borrow on the strength of their own balance sheets.

Summary 3–1. Methods of Financing Oil Projects

- In the 1990s, international oil companies (IOCs) became more financially stringent in regard to their investment decisions in developing countries. Furthermore, the governments of developing countries reduced their financial support for petroleum development and supply.
- As a result, IOCs became much more selective in undertaking investments and sought to share project risks by involving other parties, particularly local partners. These preferences led to more complex financing arrangements addressing the constraints of country limits or weaker (or less creditworthy) partners.

- Today, IOCs are in a more comfortable financial position, but the culture of investment finance has changed. Straightforward corporate financing is no longer fully applicable. Rather, various sources of funds and guarantee instruments are employed.

FINANCING POWER PROJECTS

Before the 1950s, power supply in most developing countries was run by small private companies that served specific cities and locations. In the 1950s and 1960s, the governments of these countries undertook extensive electrification programs intended to provide electricity to all social and economic groups. Along with the implementation of these programs, power grids were interconnected (for the most part) to take advantage of economies of scale in power supply. Ownership of power supply facilities was also consolidated into national (or state) utilities. These state-owned utilities, which were directly or indirectly run by their corresponding governments, became responsible for power generation, transmission, and distribution. Power supply came to be considered a socioeconomic and strategic matter with natural monopoly characteristics. Private sector participation was not deemed appropriate. Indeed, most countries had legal barriers to private sector entry into the power sector. (See fig. 3–2.)

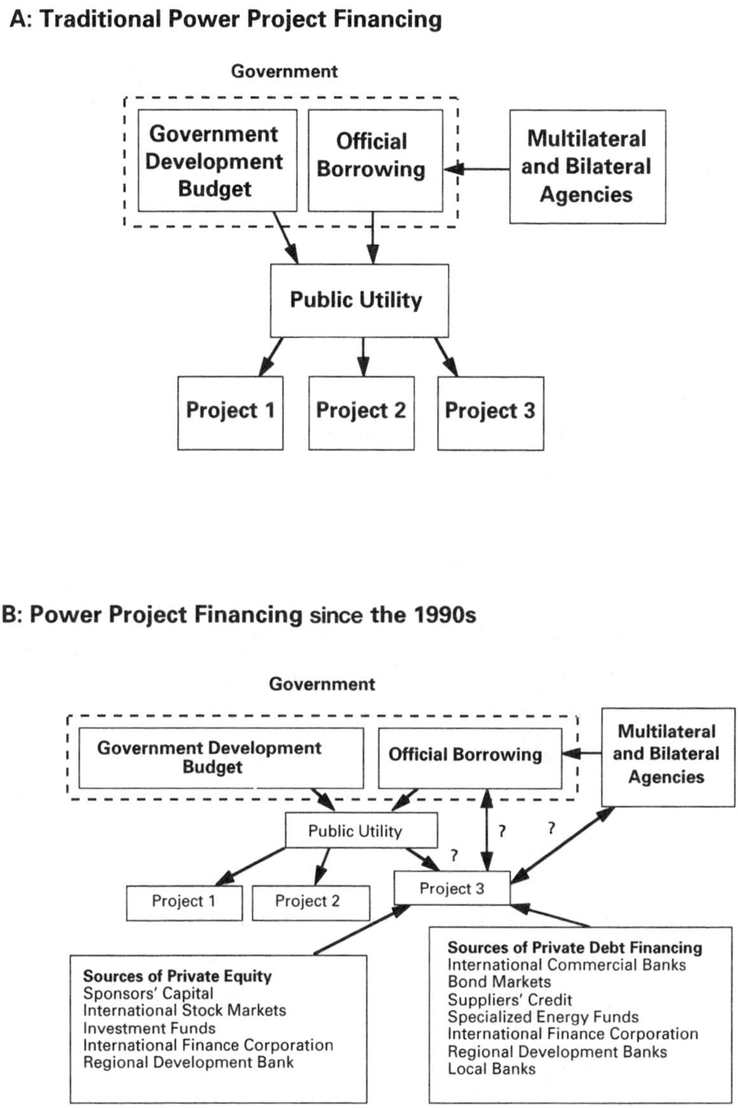

Fig. 3–2. Changes in methods of financing power projects

During this period, power project financing in developing countries involved the raising of funds by public utilities through government development budgets or through official borrowing. Official borrowing was financed by multilateral and bilateral agencies and some commercial sources. Generally, the public utility would borrow funds with government backing, usually in the form of a government counterguarantee.

Summary 3–2. Methods of Financing Power Projects

- For power projects, the traditional source of financing has been the government budget or government-sponsored borrowing.
- The inability or unwillingness of governments to continue supporting power sector investments has (1) allowed and facilitated the entry of private sector investment and (2) encouraged public utilities to borrow from commercial sources and even to raise equity on international and domestic capital markets.
- Private investment in the power sector has been less than expected and mostly in generation. The emerging trends toward corporatization and toward privatization of power systems are resulting in some private investor participation in transmission and, even more so, in distribution. However, financing for these facilities is expected to remain corporate based, because most new investments are aimed at expanding existing networks and are financially intertwined with previously built assets.

In the 1990s, the governments of most developing countries decided to limit their involvement and obligations to fund the power sector by allowing the private sector to undertake part of the required investments. Governments also encouraged state public utilities to raise private equity and finance. As a result, a number of power plants were financed with private funds on a project basis, although many power plants were still built by state utilities. Private financing of power transmission and distribution was limited to exceptional cases, where these systems have been privatized. Such projects were financed mostly on a corporate basis because most new projects are expansions of existing networks and are intertwined with existing facilities.

After the sharp decline of the U.S. equity markets in 2000–2002, the so-called strategic investors faced severe financial constraints and limited their involvement in developing countries. Since then, private sector investment in the electricity sector of these countries has become more stringent. Public sector ownership and investment in the power sector is expected to remain dominant while private sector participation is still encouraged, where suitable. However, the methods of financing power investments have become more sophisticated. Government-owned utilities tap various sources of finance. Private companies utilize innovative methods of recourse and nonrecourse funding. Finally, there are a host of new arrangements that bring together public and private resources to fund new power projects.

Financing Natural Gas Projects

Financing of natural gas projects is similar in some ways to financing of oil projects and in other ways to financing of power projects. In most developing countries, transmission and distribution of natural gas have been viewed as constituting a natural monopoly and treated like the power sector. Hence, gas transmission and distribution investments have historically been undertaken by state entities. However, upstream developments have been modeled after those of the oil sector. It has therefore been acceptable to leave upstream gas development to private companies or to joint ventures between the public and private sectors.

Factors similar to those that influence the power sector have caused many developing countries to plan to corporatize and eventually privatize their gas utilities. After privatization, investment requirements for gas systems will logically have to be funded by private sources. In certain cases—for example, when a main transmission pipeline or a new distribution network needs to be built—project-based funding will likely be possible. However, in other cases, when investments are needed to expand existing networks, financing will probably have to remain corporate based.

Summary 3–3. Methods of Financing Gas Projects

- Financing of gas projects has some similarities to financing of oil and power projects. Transmission and distribution of gas are viewed together as a natural monopoly and treated like the power sector. Upstream development is modeled after the oil sector and mostly left to private companies or joint ventures between private and public entities.

- In the event that gas transmission and distribution systems are privatized, new investments will likely be funded through corporate-based, rather than project-based, borrowing.

Sources of Funds

Analysis of the methods of financing oil, gas, and power projects indicates that

- The role of the public sector will remain important in the foreseeable future. Therefore, public and private investments will complement each other for most energy projects.
- Public (i.e., state-sponsored) projects will not be purely public because most state entities will be encouraged to borrow from or even to mobilize equity from commercial sources.
- Private sector–sponsored projects will not be purely private because most sponsors will receive assistance from governments and from official development agencies in securing debt financing and even in mobilizing equity.
- Project-based financing will not be purely on a nonrecourse basis because sponsors will likely continue to discover that some type of corporate or even government guarantee can help in raising equity and debt finance at substantially lower costs.

For the aforementioned reasons and because of the large size of most energy projects, a project planner should at least consider all types of funds, whether the project sponsor is a state entity, a project company, or a well-established corporation. These sources of financing include

- Multilateral development institutions, which were traditionally supporters of state entities but have now introduced a variety of facilities to assist private sector investment and finance. Multilaterals include global institutions, such as the World Bank and the IFC, as well as several regional development institutions that have the same objectives as the World Bank but concentrate on specific regions. The contribution of multilaterals to investment in and financing of energy projects does not exceed $15 billion annually, compared with overall annual investment requirements of more than $150 billion. Despite their relatively small contribution to the total financial requirements of the energy sector, the multilaterals' participation in any project is considered quite important because of the comfort it provides to other participating investors and financiers.
- Bilateral agencies, such as the export-import banks of industrialized countries, which traditionally lent to the state entities of developing countries to promote supply of locally produced equipment but are now willing to lend to both state and private entities under certain conditions. Bilateral agencies provide finance in the forms of suppliers' credit, buyers' credit, and guarantees. However, their assistance is often tied to their own national interests.
- Commercial finance, which encompasses many sources and instruments. Commercial sources of equity include stock markets and specialized energy funds. Sources of

debt financing include commercial banks and long-term institutional providers (of debt and equity); bond markets; and, again, specialized energy funds.

- Ad hoc sources of financing, which depend on the project and country. For example, governments of developing countries are still important sources of support for energy projects, particularly gas infrastructure and power systems. Governments' roles as direct financial contributors will likely continue to decline. Governments may, however, continue to provide subsidized energy, tax holidays, and exemptions from custom duties. Other ad hoc sources of finance include (1) project contractors, who are sometimes willing to contribute to investments and finance in order to win business; (2) sources of equipment, which are usually large multinational companies with substantial resources; and (3) purchasers of output.

Summary 3–4. Sources of Energy Project Financing

Because of market circumstances and the large size of most energy projects, a project sponsor—whether a state entity, a well-established corporation, or a new project company—must consider all sources of financing, including

- Multilateral development institutions
- Bilateral agencies, such as the export-import banks of industrialized countries
- Commercial banks and institutional providers, equity markets, bonds, and specialized energy funds

- Ad hoc sources, such as equipment suppliers' credit, project contractors' financial contributions, and equity and debt financing by purchasers of project output

4 Accessing Support From Global Multilateral Institutions

International development institutions have their origins in a gathering of delegates from 44 countries in 1944, who met to design a framework for economic cooperation following World War II. The gathering, known as the United Nations Monetary and Financial Conference, was held in Bretton Woods, New Hampshire, and resulted in the creation of two institutions: the International Monetary Fund (IMF) and the International Bank for Reconstruction and Development (IBRD), which is also known as the World Bank. The IMF was created to monitor the macroeconomic and monetary policies of member countries, to ensure that the misguided, nationalistic monetary policies of the 1930s did not reappear in the postwar period. The World Bank was initially designed to provide financing to rebuild roads, communication links, power systems, and other building blocks of the economies of war-torn Europe.

Comprehending the roles and involvement of the IMF and the World Bank is essential for project financing in developing countries because

- Both institutions have great influence on the business environments of developing countries.
- Direct financial support from these institutions significantly affects the availability of funds from other sources of financing for energy projects.

- The World Bank and its affiliates provide direct support in the form of equity, loans, and guarantees for the financing of oil, gas, and power projects.

The IMF and the World Bank are nonprofit international organizations that belong to the governments of the world. As noted, the original Bretton Woods conference was attended by representatives of 44 countries, the largest of which were the United States, the United Kingdom, France, China, India, and the Soviet Union. The nations attending the conference, except the Soviet Union, joined the membership of the IMF and World Bank. Over the years, many other countries joined, and 15 former Soviet Republics became members of the two institutions in 1991–92. The total membership in 2007 is 185 countries.

Summary 4–1. Origins and Significance of International Development Institutions

- The major international development institutions originated from the Bretton Woods conference in 1944. The conference resulted in creation of the International Monetary Fund (IMF) and the International Bank for Reconstruction and Development (IBRD), also known as the World Bank.
- Comprehending the roles of the IMF and the IBRD is essential for project financing because both institutions have great influence on a project's business environment. Their financial assistance significantly affects availability of

funds from other sources. In addition, the IBRD and its affiliates provide direct support for the financing of oil, gas, and power projects.

- The IMF and the IBRD are owned by the governments of their 185 member countries (2007). The share of each country is determined on the basis of the size of its economy.

The IMF and the World Bank are owned by their member countries. The share of each country is determined by a formula based mainly on the economic capacity of that country as compared with the capacities of the other members. The 12 largest shareholders of the World Bank are the United States (16.39%), Japan (7.86%), Germany (4.49%), France (4.3%), the United Kingdom (4.3%), Saudi Arabia (2.78%), India (2.78%), the Russian Federation (2.78%), Italy (2.78%), China (2.78%), Canada (2.78%), and the Netherlands (2.21%). A country holds voting power on the board of each institution in proportion to its shareholding.

The IMF and the World Bank are sister organizations. Their headquarters are located in Washington, DC. The World Bank is composed of two unique development institutions: the IBRD and the International Development Association (IDA). The World Bank has three affiliates: the International Finance Corporation (IFC), the Multilateral Investment Guarantee Agency (MIGA), and the International Centre for Settlement of Investment Disputes (ICSID). In addition, the World Bank houses several other important programs, such as the Global Environment Facility (GEF), the Carbon Finance Facility, and the Energy Sector Management Assistance Programme (ESMAP).

The International Monetary Fund (IMF)

Although the IMF is not directly involved with project finance, it nonetheless has an impact on energy projects in developing countries. Its role, its influence, and particularly its position vis-à-vis governments of developing countries all affect their ability to address the challenge of attracting potential financiers for energy projects.

The IMF's function at the time of its creation was to maintain the par-value system of fixed but adjustable exchange rates. It was intended to oversee the economic policies of its members and to use its financial resources to help them through periods of adjustment when extreme balance-of-payments difficulties arose. It played this role for two and a half decades, during which its jurisdiction included both industrialized and developing countries.

In the 1970s, however, major industrial countries abandoned the par-value system, and as a result, the IMF's core mission of stabilizing the international monetary system disappeared. Subsequently, the IMF adjusted its function and became an important player in recycling petrodollars in the 1970s, managing debt crises in the 1980s, and transforming former communist countries in the 1990s.

The IMF operates in two distinct ways. First, it tracks its members' monetary and fiscal policies, which can influence governments' ability to finance imports and exports. During regular consultations with government officials, the IMF recommends policy changes to correct current problems or head off future ones. Second, the IMF provides loans to member countries that have short-term problems in meeting foreign payment requirements. These loans are normally based on tough policy conditions that call for establishment of sound fiscal and monetary parameters.

Particularly in the granting of loans, the IMF's policy interventions can extend to tax, tariff, and pricing issues, which are very important from the point of view of financing energy projects. For example, in developing countries where energy prices are kept artificially low, the IMF normally pressures governments to move toward cost-based and market-related pricing of power, petroleum products, and natural gas. Such moves usually increase the commercial viability of projects in these sectors. The IMF further influences project financing when its position conflicts with government fiscal and monetary policies. When this happens, the country's credibility suffers significantly, which adds to the challenge of financing energy projects within its borders.

Summary 4–2. The International Monetary Fund

- The International Monetary Fund (IMF) oversees the economic policies of its member countries (rich and poor) and uses its financial resources to help them through periods of adjustment.
- During its regular consultations with government officials, the IMF encourages them to take remedial action to correct problems such as inflation, unemployment, and balance-of-payments deficit.
- When providing a loan, the IMF pushes for tough policy actions aimed at establishing sound fiscal and monetary parameters.

- The IMF's policy interventions extend to taxes, tariffs, and energy pricing, which are important from the point of view of financing oil, gas, and power projects. The IMF's overall attitude toward a country affects financiers' perception of the country's risks.

The International Bank for Reconstruction and Development (IBRD)

The IBRD is the original and main component of the World Bank Group. (The terms *IBRD* and *World Bank* are used interchangeably, and the term *World Bank Group* refers to the IBRD, IDA, the IFC, MIGA, and the ICSID.) When the IBRD was founded, in 1944, it operated on the principle that European countries would be short of foreign exchange for reconstruction yet would be insufficiently creditworthy to borrow commercially. As an official multilateral institution whose share capital was owned by countries in proportion to the size of their economies, the World Bank would serve as an intermediary by borrowing in world capital markets and lending to member countries in need of foreign capital. The first loans awarded by the World Bank helped finance the reconstruction of the war-ravaged economies of Western Europe. However, very soon the mission of the World Bank was extended to economic development (in addition to reconstruction) of member countries.

The objectives of the World Bank are to help reduce poverty and to foster economic development. Although these objectives have remained the same, the philosophy and the composition of the World Bank's activities have changed markedly over time. For two to three decades, the World Bank's main activity was financing investments in infrastructure, social programs, and economic activities. The decision to finance a project depended on the soundness of the project and its return as compared with alternatives and cost of finance. Projects included roads, power plants, schools, and irrigation networks, which were viewed as essential for economic and social development.

Over the years, the composition of the World Bank's activities has changed in response to changes in the world economic and political climate, the influence of stronger shareholders, and the learning process involved in economic development of member countries. In the 1960s, the World Bank reached the conclusion that providing financing does not, by itself, lead to success if the institutions responsible for implementation and operation of a project are not competent. Accordingly, most projects since then have included a component for institution building and management training. Many of today's strong power utilities were created or reinforced through institution-strengthening components of World Bank loans.

In the 1970s, the conviction was that a project would fail unless it was operated not only by a competent institution but also in the right economic environment. Attention was then extended to the correction of pricing and other policy matters affecting the business environment. This meant that a project financed by the World Bank included components for institution building and conditions or action plans to adjust prices, taxes, subsidies, and so on. In the meantime, external shocks resulting from the oil crises of the 1970s added substantial instability to the business environment. The World Bank was then drawn into fixing the economic environment not only through its project lending but also through structural adjustment loans. These latter loans provided a country with balance-of-payment support—that is, foreign exchange that could be used by a government for almost any need (excluding military spending). In return, the government was required to commit to a specified list of actions (e.g., adjusting exchange rates and removing subsidies) considered as necessary for the creation of a healthy economic environment.

Summary 4–3. The Objectives of the World Bank

- The International Bank for Reconstruction and Development (IBRD) is also referred to as the World Bank. Altogether, the term *World Bank Group* refers to the IBRD, the International Development Association (IDA), the International Finance Corporation (IFC), the Multilateral Investment Guarantee Agency (MIGA), and the International Centre for Settlement of Investment Disputes (ICSID).
- The World Bank's mission was initially to reconstruct Europe after World War II, but this was extended to cover economic and social development throughout the world.

- The World Bank's objectives of reducing poverty and stimulating economic growth have remained the same, but its philosophy and the composition of its activities have changed markedly over time.
- In its early years, the World Bank financed projects that were sound and useful for economic and social development. Today, a host of additional considerations, including pricing issues, institutional and regulatory matters, encouragement of private sector participation, and mitigation of environmental risks, come into play when a loan is granted.

In the 1980s and 1990s, the World Bank shifted its emphasis to reducing the role of the public sector and encouraging moves toward a free-market system. The World Bank's objectives are still to reduce poverty and to foster economic development. However, it now tries to achieve these objectives by creating the right business environment, in which the private sector can take the lead role in formulating, implementing, and operating most projects. In addition, the World Bank has become quite conscious of the environmental impacts of development projects. Thus, to make a project attractive to the World Bank, a project sponsor has to demonstrate how the project and the involvement of the World Bank would improve the business environment, facilitate private sector participation, or mitigate environmental problems. Because World Bank officials consider their agency the lending source of last resort, demonstration that a project would not materialize without assistance from the World Bank is also helpful in acquiring funding.

The World Bank annually lends about $20 billion, of which $2–3 billion is allocated to energy projects. The World Bank's involvement in the power sector has remained relatively stable over time. However, its lending to the oil and gas sector has fluctuated substantially, primarily because of external factors. Thus, lending to the oil and gas sector became substantial in the late 1970s and the early 1980s, when security of petroleum supply had become an international concern. The World Bank's objective was to help developing countries explore and develop indigenous petroleum resources. By 1983, its lending to the oil and gas sector had reached $1 billion per year. However, private oil companies perceived that the World Bank was preempting activities of the private sector and exerted pressure on the World Bank to limit its activities in the sector. As a result of this controversy and of the collapse of energy prices in 1986, the World Bank's lending to the oil and gas sector dropped sharply in the second half of the 1980s. Through the 1990s, however, the World Bank had become quite active in the oil and gas sector as a facilitator of private sector participation. Furthermore, the World Bank's activity had become helpful in natural gas development, which had emerged as a major area of investment and finance in many client countries.

The World Bank considers renewable energy as an area of priority and is particularly supportive of such projects, provided that they are economically and technically viable. The emphasis on renewable energy has increased in recent years in relation to international concern for climate change and also availability of targeted grants and soft loans that the World Bank can mobilize to help renewable projects.

The World Bank needs a governmental guarantee of repayment of its loans. Therefore, the World Bank's financial assistance is not directly extended to private financing of a project. However, the World Bank's involvement in a project helps and encourages private sector participation in projects in many ways:

- In joint ventures by the private and public sectors, the World Bank finances the equity or debt share of the public sector in the project.
- World Bank participation in a project often creates comfort for other financiers and investors to enter into the project.
- The World Bank can extend guarantees for political risk, foreign exchange convertibility, and so on.
- The World Bank's influence on or closeness to the host government helps facilitate legal and administrative preparation of the project package.

Summary 4–4. The World Bank's Lending to the Energy Sector

- The World Bank annually lends about $20 billion, of which $2–3 billion is allocated to the energy sector.
- The World Bank requires a government guarantee for repayment of its loans. Thus, its loans are normally made to governments and government affiliates.
- The World Bank assists private sector financing of projects through financing of public sector shares in joint ventures, through mobilization of other financiers, and through

its guarantee instruments. In certain cases, arrangements have been made for World Bank loans to support private investments through government-affiliated intermediaries.

How a loan is processed

An overview of the World Bank's loan-processing procedure is useful because it shows how the World Bank interacts with its borrowing countries and because the World Bank's procedures are very similar to those of other multilateral institutions. The World Bank has a formal loan application process that includes project identification, project preparation, project appraisal, loan negotiations, and a presentation to the Board of Executive Directors of the World Bank. The first phase, identification, is clearly essential for the project to be brought into the lending pipeline. Although only governments can propose projects for financial assistance, identification can come from several sources, including background work of the World Bank staff, of the United Nations agencies, or of private sponsors. In the energy sector, many projects are identified during the occasional studies of sectoral policy and investment strategies that the World Bank conducts for its client countries. ESMAP, a joint United Nations/World Bank effort, often reviews energy sector issues and identifies energy projects for financing by the World Bank and other financiers. At the identification stage, details of the project are not known, but the project must appeal to the World Bank objectives in order to be processed.

Project preparation begins once the World Bank and the government agree on the project objectives. The borrower is normally a public company—for example, the state power or gas utility or the NOC. The government is the guarantor but also the decision maker about how much the country should borrow and how the borrowed funds should be allocated to different sectors. Preparation is the responsibility of the borrower, although the World Bank often provides substantial assistance or at least guidance. Project preparation entails development of an idea into a detailed proposal that considers all aspects of the project—technical, economic, financial, social, institutional, and environmental. Most borrowing entities recruit international consulting firms to assist them in project preparation. At this stage, the presentation and packaging of the project are very important. The technical and economic soundness of the proposal should be firmly demonstrated by comparing the project with its alternatives. Other external benefits should also be explained and shown to satisfy the major objectives of the World Bank.

After preparatory work has been completed, a project is considered ready for appraisal. The appraisal is viewed as an independent assessment of the project by the staff of the World Bank. Its assessment includes technical, economic, financial, institutional, and environmental considerations. If project preparation is done within the framework discussed previously, the appraisal process is likely to go smoothly. Questions that an appraisal mission seeks to address are usually the same as the items described previously. Results of the appraisal work are included in an appraisal document for the project. This report is used to prepare the legal documents, usually a loan agreement or a development financing (development credit) agreement between the World Bank and the government, as well as a project agreement between the bank and the borrowing entity.

The appraisal document and the legal documents are discussed in a negotiation session between the World Bank staff, representatives of the borrowing entity, and the government. Through a give-and-take process, the World Bank and the borrower review all the issues that have arisen during preparation and appraisal. They then agree on an action plan to resolve the issues. For example, if electricity prices in the country are too low for commercial viability, the World Bank and the borrower would agree on an action plan for a gradual increase in power prices. The negotiation agreements normally include covenants to improve the institutional and regulatory framework of the energy sector, to move toward market operations, and to improve environmental and safety aspects of the energy sector. Many of these covenants are intended to have broader *development impact*—that is, they go beyond the formal scope of the project.

Summary 4–5. How to Make a Project Attractive to the World Bank

- The sponsor should demonstrate that the project—and the World Bank's involvement therein—would improve the business environment or facilitate private sector involvement or environmental remediation. It is also useful to show that the project would not materialize without the World Bank's participation.
- Preparing an attractive project package should include demonstration of its technical and environmental soundness, economic

desirability, and safe design, as well as the institutional and financial competence of the borrower.

- Many of the World Bank's loan conditions go beyond the formal scope of the project. This is what development impact is about, at least from the point of view of the World Bank.
- The final stage of loan processing is the presentation to the Board of Executive Directors of the World Bank.

The appraisal document and the legal documents are reviewed by board members for approval. Board members function in continuous session at the World Bank's headquarters in Washington, DC, and meet as often as the business of the World Bank requires. The board consists of 24 executive directors. The largest shareholders appoint one executive director; other countries are grouped in constituencies, each represented by an executive director elected by a country or a group of countries. For example, the executive directors for the United States, Japan, Germany, France, the United Kingdom, China, Russia, and Saudi Arabia represent one country each, whereas a single director speaks for 24 Francophone African countries. The country groups are formed more or less along geographic lines. However, members themselves decide how they will be grouped, and each director carries voting power equal to the sum of the capital shares of the countries in the corresponding group.

Board approval of a loan is based on the merits of the project and the strategic and political considerations of board members. Normally, by the time a project reaches the board session, the expected concerns have already been addressed, and board approval is quite likely.

Summary 4–6. How the World Bank Processes Loans

- The World Bank project cycle includes project identification, project preparation, project appraisal, loan negotiations, and board presentation.
- The Board of Executive Directors of the World Bank consists of 24 executive directors, who represent the governments of the 185 member countries.
- Board approval of a loan is based not only on the merits of the project but also on the strategic and political agendas of board members.

THE WORLD BANK'S GUARANTEE INSTRUMENT

Since the debt crisis of the 1980s and the ensuing retreat of commercial banks from financing of projects in developing countries, the World Bank has been experimenting with a number of guarantee instruments to encourage lending by commercial financiers. Until the early 1990s, these experiments were limited in scope and success and isolated from the main core of the institution's operation. In the mid-1990s, however, the World Bank decided to mainstream its guarantee operations.

Within the new framework, guarantees are issued to private lenders to cover risks arising from nonperformance of sovereign contractual obligations or from force majeure aspects of a project. This instrument is called a *partial guarantee,* implying that private lenders still bear other project risks.

The World Bank partial-guarantee instrument is especially helpful for private sector projects, the financial viability of which hinges on contractual arrangements with a number of state entities. The instrument can guarantee against defaults in sovereign contractual obligations, such as

- Government's adherence to an agreed tariff formula
- Delivery of inputs (e.g., fuel supplied to a private power plant)
- Payment for outputs (e.g., power supplied by an independent power plant to a state public utility)
- Compensation for project delays or interruptions caused by government actions or political events

In addition, transfer risks for investors and lenders may arise because of constraints in availability of foreign exchange. The World Bank's guarantee can cover the risks of converting foreign exchange.

The World Bank guarantee instrument is designed to help developing countries gain access to a wider range of private lenders and to borrow at lower costs. The World Bank's partial guarantee is triggered when a government does not comply with one or more of the obligations stipulated in agreements with project sponsors, which constitutes a default in debt service to lenders.

Unlike MIGA, the World Bank does not guarantee equity capital. The World Bank also does not guarantee loans from other official multilateral financial institutions or export credit agencies. The World Bank guarantee requires a counterguarantee from the government. Under a partial guarantee, up to 100% of the principal and interest can be covered. The World Bank charges two fees for the guarantee cover: a standby fee and a guarantee fee. These fees are charged either to the borrower or to the lender. The fees are determined on a case-by-case basis.

WHERE DOES THE WORLD BANK OBTAIN ITS FUNDS?

Like any financial institution, the World Bank has its own capital base, which comes from its member governments. However, when member governments buy shares, they pay in only a small portion of the value of each share. The unpaid balance is on call should the World Bank be unable to pay its creditors—something that has never happened. Thus, the paid-in capital is not significant when compared with the World Bank's outstanding loans of $50 billion.

The World Bank raises most of its money on the world's financial markets. It sells bonds and other debt securities to pension funds, insurance companies, corporations, other banks, and individuals around the world. Because the World Bank has the backing of almost all the governments of the world and exercises cautious financial management, its bonds have the highest rating, which enables the World Bank to borrow on good terms. This advantageous position and the World Bank's nonprofit status make its loan rates better than those on commercial loans. World Bank loans, however, are not soft, subsidized, or cheap.

There is widespread public confusion about the differences between loans from the World Bank and loans from its affiliate, IDA. By contrast to World Bank loans, IDA loans are soft. Moreover, these are part of a wholly different process, which is explained in the next section. Finally, World Bank (IBRD) loans are nonsubsidized, whereas IDA loans are subsidized.

International Development Association (IDA)

IDA was established in 1960 as an affiliate of the World Bank to provide assistance to the world's poorest countries. Currently, 82 countries are eligible for IDA financing. IDA provides interest-free loans—called *credits*—to countries with annual per capita income of less than $1,065. Many rich member countries (e.g., the United States, Japan, Germany, and the United Kingdom) and middle-income members (e.g., Brazil, Hungary, Korea, and Mexico) make financial contributions, which are made available to poor members. Thirty-two countries that were once recipients of IDA funds "graduated" from IDA (although some of them reentered the recipient status), and two of these—Korea and Turkey—became IDA donors. IDA's resources are replenished every three years. Originally, grants were designed so that each donor's contribution was proportional to its capital in the World Bank. Subsequently, donor contributions to IDA replenishments have been determined through negotiations taking into account a number of economic criteria, such as gross national product (GNP), per capita income, and trade-related indicators.

As noted previously, there is widespread public confusion about the distinctions between IDA and the World Bank. IDA is, in many respects, indistinguishable from the World Bank. Both institutions finance development projects and aim at reducing poverty. IDA has the same staff as the World Bank, and the president of the World Bank is also the president of IDA. However, each institution has separate articles of agreement, different provisions for paying in capital subscriptions, different voting structures, and separate financial resources.

The fundamental difference between the two institutions is the way they obtain funds and the terms on which they lend to developing countries. The World Bank raises most of its funds on the world's financial markets and lends to developing countries at interest rates that cover the cost of its own borrowing. By contrast, IDA provides the world's poorest countries with interest-free credits. Because of its highly concessional loan terms, IDA cannot raise funds in capital markets. Instead, most of its resources come from contributions by donor governments.

IDA's central objective is the reduction of poverty, against which it takes a twofold approach:

- Supporting a pattern of economic growth that will provide efficient employment and income opportunities for the poor while avoiding both subsidies for capital-intensive industries and damaging the environment
- Supporting investment in people in primary health, education, and nutrition with the objective of enabling the poor to participate in economic growth

Based on this approach, IDA provides credits to countries that are in serious poverty but show a determination to take the policy actions required for sustainable development.

IDA's annual lending is \$5–7 billion. Most IDA credits are extended for specific investments, although about one-quarter of lending supports structural adjustment programs. Because of its emphasis on the reduction of poverty, IDA pays special attention to social sectors, such as education, health, and nutrition. However, infrastructure projects, among them electricity supply systems, receive substantial portions of IDA credits. On average, 25% of IDA resources are allocated to the financing of infrastructure projects.

Like the World Bank, IDA has changed its philosophy of lending over time. Although, at the beginning, IDA saw its role as the financing of economically sound projects, it now looks for wider economic and social impact from its financing efforts. For example, in the power sector, the emphasis has shifted from building new projects to rehabilitating existing ones, along with improving pricing policy, regulatory processes, and institutional structures. IDA also has begun to emphasize expanding the role of the private sector and dealing with environmental concerns.

IDA lends to governments or government-affiliated entities. Its loans could finance equity or debt shares of a public entity in joint venture with private companies. The project cycle and preparation arrangements are similar to those of the World Bank. However, because of its favorable loan terms, IDA has greater leverage vis-à-vis its borrowers for improving the project's business environment of the energy sector as a whole.

To be of interest to IDA, a project must result in noticeable reduction in poverty. Other developmental impacts, such as facilitating private sector participation and reducing environmental damage, are also important to IDA.

In addition to providing direct support to poor countries, IDA plays an important role in mobilizing and coordinating aid from other multilaterals and from individual donor countries. IDA conducts systematic reviews of borrowers' public investment and expenditure programs and helps in determining investment priorities and rationalizing external aid. IDA also chairs many consultative groups, which bring donors and borrowers together. Group meetings are often the dominant forum for discussions of the economic policies, development needs, and extent of external aid by the donor community.

Summary 4–7. The International Development Association

- The International Development Association (IDA) was established in 1960 as an affiliate of the World Bank to provide assistance to the world's poorest countries.
- IDA provides the poorest countries with interest-free loans—called *credits*—to finance projects that reduce poverty.
- IDA's financial resources largely come from contributions by donor governments, including many wealthy and middle-income countries.
- IDA's annual lending is $5–7 billion.
- IDA's loan-processing procedures are similar to those of the World Bank.
- IDA plays an important role in mobilizing and coordinating aid from other donors.

The International Finance Corporation (IFC)

The IFC was established in 1956 as an affiliate of the World Bank for the purpose of promoting private enterprise in the developing world. Unlike the World Bank, which lends money only to government entities, the IFC lends to private companies and does not require guarantees of debt repayment from host-country governments. It also makes equity investments in private businesses. Quite often, project sponsors or their financial advisors or arrangers seek IFC participation not only because of the equity and financing it puts into a project but also because the IFC can mobilize additional loan and equity financing from others.

The IFC brings other investors and lenders into a project in a variety of ways. It actively seeks partners for joint ventures and encourages other lenders to participate in projects it finances. Through its syndicated loans, the IFC is able to attract large amounts of commercial bank lending to companies in developing countries. The IFC also underwrites securities issues by companies in developing countries and helps in launching country funds.

Summary 4–8. The International Finance Corporation

- The International Finance Corporation (IFC) was established in 1956 to promote private enterprise in developing countries.
- Unlike the World Bank, the IFC lends directly to private companies, takes equity in private ventures, does not require a government guarantee of debt repayment, and does not have strict procurement guidelines.
- Through its participation in a project, the IFC mobilizes other sources of investment and financing.
- The IFC assists in establishing investment funds, tapping international bond markets, and developing local capital markets.

The IFC's direct contribution to the financing of projects in developing countries is significantly enhanced through its syndication of commercial bank loans. As a multilateral institution, IFC has certain privileges; for example, it is exempt from payments of local taxes, its loans have never been rescheduled for political risk, and it has access to high-level policy makers. When the IFC syndicates a loan, it is the lender of record and brings commercial banks under its own umbrella. The commercial banks' loans thus are treated the same as IFC loans. Indeed, the IFC takes responsibility for administering the loans and collecting payments from the borrower. The IFC distributes all payments (whether received directly from the borrower or from realization of security) pro rata among the participants and itself. Thus, a default on any portion of the loan is a default to the IFC. The IFC also takes responsibility for appraising the project and coordinating the preparation of legal and contractual packages. Because the IFC takes all the lead responsibilities, commercial banks find it more convenient and comforting to participate in financing projects in developing countries.

In the area of infrastructure, the IFC's participation in project finance has been very effective in mobilizing other resources. It is estimated that for every dollar invested by the IFC, six dollars were provided by others. This is particularly important in the power sector, where the IFC has become an important player since the 1990s.

The IFC's involvement in the oil and gas sector started in the late 1970s, grew at a moderate pace during the 1980s, and surged markedly during the 1990s, as an increasing number of countries facilitated private investments in the sector. The IFC invests in all parts of the oil and gas sector, upstream and downstream, except pure exploration. Where appropriate, it is prepared to participate in and lend to unincorporated joint ventures.

The IFC has been quite innovative in the development of instruments for mobilizing debt and equity financing. Several of these instruments are of direct relevance to the financing of oil, gas, and power projects:

- The IFC's participation in build-operate-transfer (BOT) and build-own-operate (BOO) schemes has proven very useful in bringing in other investors and financiers and in resolving contractual, legal, technical, and financial issues with governments and government entities.
- The IFC has taken an active role in the creation of specialized investment funds to mobilize resources from the world's major financial centers for lending and equity investment in developing-country projects.
- The IFC assists private companies of developing countries in using international bond markets and international equity placements. The IFC helps companies design these instruments and provides loans and underwriting commitments.
- The IFC has become involved in developing local capital markets. Again, the IFC provides broad technical assistance in this area. However, it also helps in mobilizing funds for specific projects.

Since the IFC does not require (and cannot accept) government guarantees of debt repayment, the success of its investments depends on the commercial success of the projects it helps finance. As a result, the IFC's application-processing procedure is more directly concerned with the various aspects of a proposed project than with the economy, the sector, and other broad policy matters that are of concern to the World Bank and IDA. Nevertheless, the IFC's objective is not only to support private projects but also to encourage private investment and financing in the sector. It is therefore interested in improving the overall business environment and is careful not to compete with private sources of funds. Thus, an IFC appraisal pays attention to regulatory matters and the availability of other sources of funds. In addition, the IFC follows the same standards of environmental safety as does the World Bank. However, the IFC does not require borrowers to follow strict procurement guidelines but relies on good commercial practices.

In approaching the IFC for support, project sponsors should be aware that

- The IFC's interest rates and fees are determined on a commercial basis. The overall terms run from 7 to 12 years.
- The IFC will not finance more than 25% of the project cost for a greenfield project and can go slightly higher for project expansions.
- The IFC will take only a minority share in any entity.
- The IFC's shareholdings are usually treated as domestic or neutral capital for nationality ownership purposes.
- The IFC is a long-term equity investor, but when appropriate, it seeks to divest by selling its equity to local private investors either directly or through local stock exchanges.
- The IFC does not take an active role in the day-to-day management of a venture in which it participates.
- The IFC charges the project sponsors for the staff and resources it spends on preparing and appraising a project.

Summary 4–9. The International Finance Corporation's Involvement in the Energy Sector

- The International Finance Corporation (IFC) has had a presence in the oil and gas sector since the late 1970s.
- The IFC's involvement in private power projects has grown in importance since the 1990s.
- The IFC mobilizes other sources of funds in a variety of ways that are particularly important in the case of large power projects.
- The IFC's appraisal processes are directly concerned with project risks, viability, and success.

Multilateral Investment Guarantee Agency (MIGA)

MIGA was established in 1988 as an affiliate of the World Bank to encourage foreign investment in developing countries by providing investment guarantees against the risks of currency transfer, expropriation, war, civil disturbances, and breach of contract by the host government. MIGA has issued nearly 850 guarantees in 92 countries for projects totaling more than $16 billion in coverage. MIGA's gross exposure is currently about $5.5 billion. Its membership includes 170 developing and developed countries.

MIGA covers four categories of project risks, which can be purchased individually or in combination:

- *Currency transfer coverage* protects against losses arising from the investor's inability to convert local currency returns (profits, principal, interest, royalty, etc.) into foreign exchange for transfer outside host country. Coverage does not include currency devaluation.
- *Expropriation coverage* protects against partial or total loss of investment as a result of acts by the host government that may reduce or eliminate ownership or control of assets.
- *War and civil disturbance coverage* protects against losses from asset damage, destruction, or disappearance caused by politically motivated war or civil disturbance, including revolution, in the host country.
- *Breach-of-contract coverage* protects against losses arising from the host government's breach or repudiation of a contract with the investor.

MIGA's standard term is 15 years and covers both equity and loans. For equity, MIGA covers up to 90% of investment plus an additional 180% to cover earnings attributable to the investment. For loans, MIGA covers up to 90% of the loan and the interest that will accrue over the term of the loan. Rates vary according to sector, project, and type of coverage.

MIGA has established itself well in the financing community over a short period. Its application-processing procedure is simple and efficient. However, its coverage is limited to $200 million per project and $620 million per country. For certain countries, this may be a small facility in the light of the financing needs of oil, gas, and power projects. Nevertheless, incorporation of MIGA into a financing scheme facilitates mobilization of funds and guarantees from other agencies. Furthermore, MIGA may be able to provide greater coverage by syndicating insurance from other agencies.

Energy Sector Management Assistance Programme (ESMAP)

After the oil crises of the 1970s, the United Nations Development Programme (UNDP) and the World Bank decided to form a joint program for assisting developing countries in dealing with energy crises. Initially called the Energy Assessment Programme, the effort concentrated on broad sector studies of energy development issues and strategies. After completing such studies for most developing countries, the Energy Assessment Programme began emphasizing follow-up studies on specific issues such as energy investment programs and energy pricing, along with subsectoral investigations in the areas of oil, gas, and power. In keeping with this shift, the name of the program was changed to the Energy Sector Management Assistance Programme (ESMAP).

Presently, ESMAP is a joint program of the World Bank and 10 other donors, mainly European bilateral aid agencies. The Netherlands and the United Kingdom are the largest contributors. The total donor support is about $10 million per year in the form of core and noncore funds. The core funds are paid by some donors in predetermined amounts each year.

ESMAP management decides how to allocate these funds to various studies. The noncore funds are in turn categorized into three types differentiated by restrictions that donors impose on the use of funds:

- Thematic funds, restricted for use in certain themes (e.g., household and rural energy or energy efficiency)
- Country/regional funds, restricted for use in certain countries
- Project-specific funding, requiring the donor's approval on a case-by-case basis

ESMAP is managed by the World Bank on behalf of the contributing bilateral donors. Although its budget is small, ESMAP is nonetheless influential. ESMAP does not deal with detailed engineering or feasibility studies. Instead, it concentrates on resolving strategic and policy issues. While these activities require a small budget, they still have a significant impact on governments' energy policies. During recent years, ESMAP has assisted a number of developing countries in privatizing their oil, gas, and power entities. It has also helped set up the appropriate regulatory framework to attract the private sector in upstream oil and gas activities, independent power generation, and downstream petroleum and gas distribution. Also, as a matter of priority, ESMAP has provided support to promote energy efficiency and renewable energy and to increase energy access to the poor. Since each ESMAP activity is, in practice, managed by World Bank staff, the results of the activity are reviewed by a high-level decision-making audience. In addition, because of World Bank relationships with government authorities, ESMAP study teams have easier access to background policy matters than do consulting and other research groups.

Although ESMAP does not provide direct financing for project preparation, it can offer much to those who are trying to structure a project-finance package. The results of its past and ongoing activities are often relevant to various aspects of project preparation. For example, a major component of project preparation deals with projection of future regulatory changes in a country; possible adjustments to domestic petroleum, gas, and power prices; or expansions of gas and power grids. ESMAP activities address these issues on the basis of first-hand information in each country.

Most ESMAP reports are publicly available. ESMAP staff can help project developers acquire in-depth information about strategic and policy issues in the oil, gas, and power sectors and about technical and economic aspects of renewable energy.

Summary 4–10. The Energy Sector Management Assistance Programme

- The Energy Sector Management Assistance Programme (ESMAP) was established as a joint program of the United Nations Development Programme (UNDP) and the World Bank. It now has 11 sponsors, including the World Bank and 10 bilateral donors, most of which are European. ESMAP is managed by the World Bank.
- Because of ESMAP's concentration on the resolution of strategic and policy issues and because of its access, through the World Bank, to high-level government authorities, the

program has become an influential instrument in the shaping of countries' energy investment and sector strategies.

- ESMAP's past and ongoing studies contain useful information for the preparation of energy projects, especially on strategy and policy issues. They contain in-depth analysis of energy sector reform, government policy, and economic and technical aspects of oil and gas, power, and renewable energy.

The Global Environment Facility (GEF)

GEF was established in 1991 to help developing countries fund projects and programs that protect the global environment. GEF's areas of focus include biodiversity, climate change, degradation of international waters, ozone depletion, land degradation, and organic pollutants. Since its establishment, GEF has directed the bulk of resources toward biodiversity and climate change, which receive about 35% and 25%, respectively, of the total GEF resources.

In dealing with climate change, GEF provides funds for activities that reduce CO_2 and methane emissions. Fuel switching, energy efficiency, and renewable energies are all considered favorably by GEF. For instance, GEF has supported solar-based power plants in Morocco and wind power developments in Ukraine, Tunisia, and Pakistan. It has also supported biomass power generation and cogeneration in Thailand and Brazil.

The climate change agenda is closely linked to the energy sector activities. The agenda has six strategic priorities, of which three are related to the promotion of renewable energy. These span the improvement of the institutional structure (regulatory framework), innovations in/advancement of renewable energy technologies, and promotion of productive uses of renewable energy. The remaining priorities of the climate change agenda focus on market transformation programs for energy-efficient products, leveraging the financing of energy-efficiency projects and the promotion of more efficient public transport. Renewable energy and energy efficiency are of particular interest in this context because they can be applied to numerous energy projects if properly formulated and presented.

GEF grants fall into three categories depending on the size of the grant. Grants over $1 million are considered as full-size grants and are processed in a fashion similar to a World Bank loan. They go through the process of identification, preparation, appraisal, negotiations, and board presentation. For medium-size grants (up to $1 million), the process and documentation are simpler. The project goes through the same processing but does not need to be approved by the board of directors. Finally, for "enabling" activities, aimed at assisting countries in various aspects of devising their climate change agenda, GEF provides up to $450,000 on a fast-track basis.

GEF works with a number of regional institutions to help with the management and the implementation of projects. This allows GEF to delegate responsibility and gain from the expertise of these institutions. Partnerships include the

Asian Development Bank, the African Development Bank, the European Bank for Reconstruction and Development, the Inter-American Development Bank, the International Fund for Agricultural Development, the United Nations Food and Agriculture Organization, and the United Nations Industrial Development Organization.

Summary 4–11. The Global Environment Facility

- The Global Environment Facility (GEF) provides grants for studies and actual projects that reflect regional or national priorities and have favorable impact on the global environment.
- The main areas of GEF's focus are biodiversity and climate change. Energy activities represent a big portion of the climate change component.
- GEF has been particularly active in fuel switching, energy efficiency, and renewable energy.
- GEF can be tapped at a large scale, when cofinancing an energy project, or on a small scale, for demonstration projects and studies.

The Carbon Finance Facility

The concept of carbon finance has emerged in recent years in response to the opportunity that OECD countries may be able to meet their commitment to reduce greenhouse-gas emissions through projects in developing countries. The incentive for such an arrangement is that it may cost substantially less to reduce by a certain amount emissions in, for example, Indonesia than in Japan; therefore, Japan would be able to pay for some part of a suitable project in Indonesia and use the emission reduction credit to satisfy Japan's overall commitment to carbon reduction. This idea has now evolved into a host of instruments managed by the World Bank Carbon Finance Unit (CFU), which uses the money contributed by governments and companies in OECD countries to purchase project-based greenhouse-gas emissions reductions in developing countries. The emission reductions are purchased through one of the CFU's instruments and within the framework of the Kyoto protocol.

Carbon finance can contribute to the financing of an energy project in a developing country provided that the project is considered suitable—that is, its emissions reduction potential is clearly established. However, the money from carbon finance does not usually come during the project construction and implementation, but flows in during the operations phase, when the project receives an annual payment in proportion to the verified emission reduction credit. In other words, the project sells its emission reduction to increase its bankability by adding an additional revenue stream in hard currency, which improves the economics of the project and reduces the risk of financial failure.

Within the framework of carbon finance, the World Bank has established a number of funds to meet different objectives:

- Prototype Carbon Fund was the first carbon fund and is used as a learning model.
- BioCarbon Fund targets projects that conserve carbon in forest and agricultural ecosystems.
- Community Development Carbon Fund supports projects that combine community development attributes with emission reductions.
- Umbrella Carbon Facility is an aggregating facility to pool funds from existing carbon funds for the purchase of emission reductions from large projects.

There are also a number of funds that have been established through agreements between the World Bank and public and private entities of the respective countries:

- Italian Carbon Fund
- The Netherlands Clean Development Mechanism (CDM) Facility
- The Netherlands European Carbon Facility
- Danish Carbon Fund
- Spanish Carbon Fund
- Carbon Fund for Europe

When formulating an energy project in a developing country, the project sponsor should consider applying for funds from the CFU. The present CFU program amounts to about $1 billion and provides ample opportunity for the support of qualified projects. The CFU has developed a user-friendly interface that assists in the preparation of carbon documentation; CFU may also agree to pay part of the future credit in the form of up-front payment to project developers.

Summary 4–12. The Carbon Finance Facility

- The carbon finance concept and its corresponding instruments have emerged to enable countries in the Organisation for Economic Co-operation and Development (OECD) to meet their greenhouse-gas emission reduction commitments at a lower cost than can be done in their own home countries.
- There are now formal processes for channeling the funds from an entity in an OECD country to a project in a developing country and for certification of the emission reduction credit to the OECD entity.
- The World Bank Carbon Finance Unit (CFU) manages the process through a host of instruments.
- Carbon finance instruments can be used to increase economic and financial viability of certain energy projects. However, the payment is normally received during the operations phase, rather than construction phase of the project.

5 Borrowing From Regional Development Banks

Most regional development banks were created in the 1950s and 1960s with objectives similar to those of the World Bank but focused on specific regions. Also like the World Bank, the regional banks were designed to assist in reducing poverty and in promoting economic growth. Moreover, they were modeled after the World Bank with respect to procedures for project preparation and implementation.

In this chapter, four regional development banks are described: the Inter-American Development Bank (IDB), the Asian Development Bank (ADB), the African Development Bank (AfDB), and the European Bank for Reconstruction and Development (EBRD). These banks are owned by the governments of the corresponding regions and by the governments of industrialized countries. Therefore, they have a mandate and a political structure similar to that of the World Bank.

Also described in this chapter are a number of other multilateral institutions, such as the European Union (EU), the European Investment Bank (EIB), the Nordic Finance Group, the OPEC Fund, the Arab Fund, and the Arab Bank for Economic Development in Africa (Banque Arabe pour le Développement Economique en Afrique [BADEA]). Each institution has its own objective, as well as its own orientation and membership. For example, the United States is not a member in the first three institutions, and the whole set of OECD countries are not members of the remaining institutions. As such, these multilateral institutions have different political structures and orientations. Still, they are multilateral institutions with significant activities in developing countries and with some regional focus.

The Inter-American Development Bank (IDB)

The IDB is the oldest of the regional development institutions. It was established in 1959 to advance economic and social development in Latin America and the Caribbean. The IDB's membership originally included 19 Latin American countries and the United States. Over time, the membership expanded and has now reached 47 countries, of which 26 are borrowing members and the rest are mostly industrialized countries. The largest shareholder is the United States, with 30%, followed by Argentina and Brazil, with 10.75% each; Venezuela, with 5.76%; and Japan, with 5.0%.

The IDB has a subscribed capital of $101 billion. Only 4.3% of this capital is paid in by the member countries; the remaining 95.7% is callable capital that comprises guarantees by member countries to provide funds if the IDB cannot repay what it borrows on the capital markets. Therefore, in practice, the bulk of capital is obtained through IDB's own borrowing.

The IDB's annual lending is about $9 billion and covers all social and economic sectors. It lends to public entities and requires a government guarantee. A major part of the total lending (20%–25%) is allocated to the energy sector. Like the World Bank, the IDB was initially pursuing the mandate of supporting economic development in a broad sense. In the 1990s, it shifted its emphasis to private sector development and environmental protection. The IDB now has a formal strategy that identifies four specific areas in which its efforts are focused:

- Social development, particularly poverty reduction through sustainable economic development, education reforms, and promotion of social inclusion
- Competitiveness, particularly support of activities that improve countries' ability to compete in the global economy, as well as to strengthen their financial sectors and capital markets
- Integration at subregional, regional, and hemispheric levels
- Modernization of state activities, including the processes of state reform and institutional strengthening at the national and subnational levels

The elaboration of the IDB's strategy is an important achievement. It describes in a nonideological and a practical manner a comprehensive framework for the support of its member countries.

In association with the IDB's activities, the Fund for Special Operations (FSO) has $9.8 billion in paid contributions from all member countries. FSO is used for lending to the poorest countries of the region. It lends about $400 million per year. The terms of an FSO loan include a 40-year maturity, 10-year grace period, and an average interest rate below 2%.

Although there are two IDB-associated entities—the Inter-American Investment Corporation (IIC) and the Multilateral Investment Fund (MIF)—that support the private sector, the IDB itself allocates up to 10% of its loans to be made directly to private businesses without government guarantees. These operations are typically in the infrastructure area, including the energy sector. The IDB's participation in any single private sector project is limited to 25% of the project cost and a $200 million ceiling, but these limits can be exceeded in exceptional cases. The private sector loans are priced in accordance with market conditions.

As mentioned previously, the IIC and MIF are associated with the IDB and were established to support the private sector, but normally on smaller projects and smaller entities. IIC operations may take the form of loans, investment in stock capital, or guarantees. Eligible projects include new companies or the expansion, modernization, restructuring, or privatization of existing companies. The IIC finances up to 33% of project cost. It will hold no more than one-third of the share capital. It will not assume administrative or managerial duties in the companies. The IIC's total direct participation in a project is limited to $10 million. This small ceiling, imposed by the IIC's capitalization status, limits IIC involvement in most energy projects. However, because of its multilateral status, the IIC's presence in a project facilitates equity and debt financing by other sources. The IIC itself takes responsibility for arranging syndicated loans. Under such arrangements, the IIC provides a large loan, which in effect is made up of contributions from syndicating commercial banks. The IIC remains the lender of record, serves as the administrator of the loan, and conducts the project's feasibility studies. A default on any portion of the loan would be treated as a default to the IIC. In addition to the greater security for commercial bank participants, IIC syndication arrangements provide advantages such as tax exemptions and a closer working relationship with host governments. The IIC's projects in the energy sector are primarily in private power generation.

MIF was established in 1992 as a technical assistance mechanism to strengthen the environment for small and medium enterprises (SMEs). The United States and Japan are the biggest contributors to the fund. The fund has provided $950 million to approximately 660 projects. Its main activity has been in technical assistance to regulatory reform, training, and efficiency improvements of small enterprises. A recent relevant activity in the energy sector is establishment of a $5.5 million fund that supports the Central American renewable energy facility; this fund provides mezzanine financing for SMEs in renewable energy. MIF also has funded preparation work on legal and institutional reforms of the electricity, oil, and gas sectors of several countries.

Summary 5–1. The Inter-American Development Bank

- The Inter-American Development Bank (IDB) is the oldest regional development institution. Founded in 1959, it has 47 member countries, of which 19 are nonregional industrialized countries.
- The IDB's annual lending is about $9 billion, of which 20%–25% goes to the energy sector. The IDB lends to public entities and requires a government guarantee. It also has a grant program that supports project preparation, sectoral studies, and other technical assistance activities.

- The IDB has a $9.8 billion program—the Fund for Special Operations (SFO)—that provides concessional loans to very poor countries in the region.
- The IDB allocates up to 10% of its lending to direct support of the private sector—normally to infrastructure projects, including private energy operations.
- Further support of the private sector is provided through two IDB affiliates: the Inter-American Investment Corporation (IIC) and the Multilateral Investment Fund (MIF).
- The IIC provides equity and loans to private companies without a government guarantee. IIC participation is limited to $10 million per project, but it can arrange syndicates to mobilize commercial bank loans.
- MIF provides technical assistance and financial support to activities aimed at advancement of market-oriented sector reforms, project preparation, and institution building.

The Asian Development Bank (ADB)

The ADB was formed in 1966. Its membership includes 67 countries (48 from Asia and 19 from outside the region, most of which are industrialized.) The largest shareholders are the United States and Japan, with 12.8% each, followed by China and India, with 5.4% each; Australia, with 4.9%; Indonesia, with 4.6%; Canada, with 4.5%; and South Korea, with 4.3%. The ADB's ordinary capital is $54 billion. This subscribed capital consists of $3.8 billion paid-in capital and the rest as callable capital (guarantees by member countries to provide funds in the event that the ADB cannot repay funds it has borrowed on the capital markets).

The ADB has a program—the Asian Development Fund (ADF)—that provides concessionary funding to very poor countries of the region. The ADF was established in 1973 with initial contributions from the richer member countries. It has been continuously replenished and has now reached $7 billion.

The ADB's total annual lending is about $6 billion, of which 15%–20% goes to the energy sector. Lending has primarily been to traditional public projects, particularly power generation, transmission, and distribution projects. In its traditional lending, the ADB requires a government guarantee. However, since the 1980s, it has emphasized private sector participation. Support for the private sector increased in 1983 with the introduction of an equity investment facility, which allowed the ADB to make direct equity investments in private enterprises and financial institutions and to extend lines of equity to selected financial intermediaries. In 1985, the ADB further reinforced its support for the private sector by establishing a facility for direct lending to private enterprises without government guarantee.

The private sector loans are based on a fixed spread over LIBOR. The spread is determined on a case-by-case basis on assessment of credit and project risks and is set prior to loan signing.

The ADB has been very supportive of BOT and BOO projects in the power sector. For these projects, it provides equity and loans with and without government guarantee. This is particularly useful for projects in which a public entity holds a minority share. The ADB is then able to finance the public entity's equity and debt share, through a loan with government guarantee, and the private sector's equity and debt share, through direct participation.

Summary 5–2. The Asian Development Bank

- The Asian Development Bank (ADB) was founded in 1966. Its members include 48 Asian countries and 19 countries from outside the region.
- The ADB's annual lending is about $6 billion, of which 15%–20% goes to the energy sector. It traditionally lends to public entities with government guarantee but has established an equity investment facility and a private sector lending facility to provide equity and debt financing without government guarantees.
- The ADB has a $7 billion program—the Asian Development Fund—to provide concessional loans to very poor countries in the region.
- The ADB provides grant funds for technical assistance that can be utilized for certain aspects of project preparation. The availability and size of these funds are generous in comparison with resources allocated to this matter by other multilateral institutions.

The ADB also provides grants that can be used to study strategic issues and policies related to sector management and investments. Many of these grants have been used to perform studies in the energy sector, which have ranged from investigation of broad sector issues to specific preparatory work for significant investments in gas and power. The size of the grant program—about $200 million per year—is quite large compared to what is normally available through multilateral institutions. Approximately 15%–20% of this grant is used in the energy sector. The ADB also has at its disposal a substantial grant facility funded by the government of Japan and other facilities furnished by the governments of other OECD countries.

The African Development Bank (AfDB)

The AfDB was formed in 1964. Its membership includes 53 African states and 24 other nations, most of which are industrialized. The AfDB's authorized capital is $33 billion, which consists of 12.5% paid-in capital and 87.5% callable capital. The AfDB's headquarters are in Abidjan, Côte d'Ivoire; however, since 2003, the AfDB has been operating from Tunis, Tunisia, owing to security problems in Côte d'Ivoire. The *AfDB Group* refers to the AfDB itself, the African Development Fund (established in 1972), and the Nigerian Trust Fund (established in 1975); the AfDB is the most significant entity in the AfDB Group and provides about two-thirds of its total lending. The AfDB borrows on capital markets and lends to the more developed nations of Africa. Project loans have maturities of up to 20 years at interest rates that are tied to the cost at which the AfDB borrows its funds.

Summary 5–3. The African Development Bank

- The African Development Bank (AfDB) was formed in 1964. Its membership includes 53 African states and 24 other nations.
- The AfDB provides loans at rates tied to its own cost of borrowing on capital markets.
- An AfDB affiliate, the African Development Fund, receives financial contributions from nonregional member countries and provides concessional loans to the poorest African countries. Another AfDB affiliate, the Private Sector Development Unit, provides financial support to private sector projects.

The African Development Fund is a concessional fund similar in concept to IDA, of the World Bank Group. Contributions to the African Development Fund are made by the AfDB's nonregional members. The fund, managed by AfDB, provides low-cost loans to Africa's poorest nations. Loan maturities are up to 50 years, and no interest is levied except for a 0.75% service charge.

The Nigerian Trust Fund was established by the AfDB and the Nigerian government and is another part of the AfDB Group. However, its capitalization and range of activities are comparatively minor.

The AfDB Group's annual lending is approximately $4.5 billion, of which about 20% is directed to the energy sector. Like other multilaterals, the AfDB supports private sector investments. In 1991, it established the Private Sector

Development Unit (PSDU), which operates as a separate organizational entity. The PSDU mobilizes financial resources for private sector projects through term loans and equity participation. PSDU loans and investments do not require government guarantee. The terms of these loans are linked to the market. An interesting feature of the PSDU's activities is that it can lend under commercial terms to publicly owned entities without sovereign guarantee, provided that these entities are financially sound and autonomously managed.

The European Union (EU)

Creation of the EU has had a major influence on developing countries in the vicinity, as well as many others. The EU provides about €7 billion per year of direct support to developing countries in addition to many more programs of assistance by its individual member countries and associated entities and programs. In the Treaty of Rome of 1957, the six signatory countries created the European Economic Community (EEC). After merging with the European Coal and Steel Community and the European Atomic Energy Community in 1967, the EEC became the European Communities or the European Community (EC). The EU was established in 1993 with the entry into force of the Maastricht Treaty on European Union.

The aim of the EU is to integrate member economies fully. The main institutions of the EU are the Council of Ministers, the European Parliament, the Court of Justice, and the European Commission. The EU exercises supranational powers. Its membership has now expanded to include 27 countries.

The EU pays special attention to relationships with developing countries. It has signed a large number of international agreements with developing countries, involving the full range of aid and trade development instruments, and maintains a permanent dialogue with them on policy issues and cooperation. EU assistance has been extended to countries in Africa, the Caribbean, the Pacific, the southern and eastern Mediterranean areas, Latin America, Asia, Central and Eastern Europe, and the former Soviet Union (FSU).

The EU provides substantial aid to developing countries, with different programs for different regions. The European Development Fund (EDF) provides funds for African, Caribbean, and Pacific countries (ACP). The EDF consists of several instruments, including grants, risk capital, and loans to the private sector of the ACP. The first EDF was implemented back in 1954. The ninth EDF came into effect with the Cotonou Agreement in 2000, which provided a structure for the aid.

Considerable loans are also provided for projects in overseas countries and territories (OCTs). The OCTs comprise 21 countries that have a special relationship with one of the member states of the EU.

The Technical Assistance to the Commonwealth of Independent States (TACIS) program came into effect in 1991. It provides support to the countries of Eastern Europe and Central Asia and is now a component of the EuropeAid program. TACIS helps these countries to move toward market-based economies. Technical assistance is provided for infrastructure projects, and environmental protection and management of natural resources are promoted.

The Asia and Latin America (ALA) program was set up to assist countries in these regions. Asian countries such as Brunei, Burma, China, Cambodia, Indonesia, the Philippines, and Vietnam have obtained significant support recently from the EU. There are a number of programs and treaties between the Asian countries and the EU in the field of energy. For instance, the EC-ASEAN Energy Facility is a program of cooperation between the EC and the Association of Southeast Asian Nations (ASEAN), to foster partnerships in developing energy projects. The objective of the EC-ASEAN Energy Facility is to secure energy needs of ASEAN nations; its intent is also to improve the environment locally and globally.

The Euro-Mediterranean Partnership program (MEDA) was established to support Mediterranean countries. MEDA is the major financial instrument through which the EU promotes economic and social development in countries such as Morocco, Algeria, Tunisia, Egypt, Israel, Jordan, Lebanon, Syria, Turkey, and Cyprus, as well as Palestine (Gaza/West Bank).

The European Investment Bank (EIB)

The EIB is the financing institution of the EU. The members of the EIB are the member states of the EU, all 27 of which have subscribed to the EIB's capital. The main capital is provided in accordance with the economic weight within the EU, as expressed by gross domestic product (GDP) at the time of a country's accession. The top five contributors are Germany,

France, Italy, and the United Kingdom, with 16.2% each, and Spain, with 9.7%. The volume of the EIB's operations is about $25 billion a year. Over the past five years, around 9% of the EIB's funds have gone to energy projects. Although the bulk of its loans are within the EU, the EIB has been called on to participate in the EU's development cooperation activities.

Over the past 10 years, the EIB's development activities have been focused on the accession countries, all of which except Turkey, Croatia, and Macedonia have now joined the EU. In other nations, such as Russia, Ukraine, Moldova, and Belarus, the EIB has initiated support to infrastructure and energy projects. In addition, with the objective of fostering relationships between the EU and the wider community of Europe and North Africa, the EIB assists Mediterranean nonmember countries: Algeria, Egypt, Jordan, Lebanon, Morocco, Syria, and Tunisia, as well as Palestine (Gaza/West Bank).

The EIB's involvement in Mediterranean nonmember countries has increased particularly since the 2002 EIB's establishment of the Facility for Euro-Mediterranean Investment and Partnership (FEMIP). The FEMIP emphasizes projects aimed at energy efficiency and renewable energy. It also finances projects in the power and gas sector—in particular, those that facilitate energy trade. A well-known example is the trans-Mediterranean pipeline, a regional transport network that carries gas from Algeria to Spain and Italy. Other examples include gas development projects in Algeria and Egypt.

The EIB's support of Africa has focused on essential projects in the energy sector. For example, it provided €43 million to Kenya in 2005 to support the electricity grid network and €50 million to Ethiopia to support the construction of a hydroelectric power plant on the Gilgel Gibe River in 2005.

The EIB's support of the energy sector has been extended to certain ALA countries. These include power transmission and distribution projects in Brazil and a hydroelectric power project in Laos.

The European Bank for Reconstruction and Development (EBRD)

The EBRD was established in 1991 to foster the transition of the economies of Central and Eastern European nations (and the FSU countries) toward open, market-oriented operations and to promote private initiative in Central and Eastern European countries. Membership includes 60 countries plus the EU and the EIB. Subscribed capital amounts to €20 billion (€5 billion paid-in capital and €15 billion callable capital). The United States subscribes to the largest amount, at 10% of capital, followed by the United Kingdom, Germany, Japan, France, and Italy, with 8.5% each. The EBRD also borrows on international capital markets to meet member countries' demand for debt and equity funds.

The EBRD offers loans with and without government guarantee. The loans with sovereign guarantee are made to public sector entities. The loans without sovereign guarantee are extended to the private sector and, under certain conditions, to public entities that are managed autonomously. In all of its operations, the EBRD attempts to foster the role of the private sector. It is also required to lend only to countries that are committed to democratic principles.

Summary 5–4. The European Investment Bank and the European Bank for Reconstruction and Development

The European Investment Bank (EIB) and the European Bank for Reconstruction and Development (EBRD), although related, were formed with entirely different objectives:

- The EIB was set up in 1958 to finance investments in the 6 member countries of the European Economic Community (EEC), which in 1967 became the European Community (EC) and in 1993 the European Union (EU), which now has 27 member countries. The EIB has on occasion provided support to projects in developing countries. From an annual lending of about $25 billion, about 10% normally flows to developing countries.
- The EBRD was established in 1991 to assist with the transition of Central and Eastern European countries and the countries of the former Soviet Union (FSU) to market-oriented economies. Its membership includes 60 countries plus the EU and the EIB. The EBRD provides loans with and without sovereign guarantee. Its loans to the private sector are made without government guarantee. It can also invest in private projects.

THE NORDIC FINANCE GROUP

The Nordic Finance Group consists of four multilateral financial institutions: the Nordic Investment Bank (NIB), the Nordic Development Fund (NDF), the Nordic Environmental Finance Corporation (NEFCO), and the Nordic Project Fund (NOPEF). Each of these entities has components that deal with the investment activities in developing countries.

The NIB was established in 1975 by the five Nordic countries—Denmark, Finland, Iceland, Norway, and Sweden—as a multilateral financial institution to finance investments of interest to its members. The membership of the NIB was extended in 2005 to include Estonia, Latvia, and Lithuania. While the bulk of its activities are focused in its member countries, the NIB attempts to expand its presence in the neighboring areas and in other countries where a mutual interest is identified. The NIB's activities in emerging countries include projects in Asia, the Middle East, Central and Eastern Europe, Latin America, and Africa. In most of these countries, the NIB attempts to support projects that promote "internalization" of the business. Energy trade projects receive a high priority.

The NDF was established in 1989 as a multilateral financial institution funded by the aid budgets of the Nordic countries. The NDF provides long-term credits to low-income and lower-middle-income developing countries on concessional terms to promote social and economic development. The NDF engages only in cofinancing and builds its portfolio on operations initiated by other institutions. It has a limited staff. The NDF's cofinancing partners are the World Bank, the IFC, the IDB, the

AfDB, the ADB, and the NIB. Energy is included among the sectors of interest of the NDF, and countries of interest are low-income developing nations, primarily in Africa and Asia.

The NDF's credits are granted on the same terms to all eligible countries. The terms are concessional—that is, interest free—with an annual service charge of 0.75%. Credits have 40-year maturities, including a 10-year grace period. Credits for public sector projects are in the range of $1.3–13 million.

NDF funds are primarily intended for procurement from Nordic sources, under Nordic competitive-bidding procedures. When consultant services are financed, the NDF encourages cooperation between Nordic and local consultants and sometimes finances certain local services.

NEFCO is an international financial institution that supports projects with positive environmental contributions to Central and Eastern European countries. NEFCO supports projects through equity investment, loans, and guarantees.

NOPEF grants Nordic countries favorable loans and grants for feasibility studies aimed at export promotion and internationalization projects. Its target countries are Turkey and the countries of Eastern Europe and the former Soviet Union. The priority of NOPEF is to work with small businesses.

The Islamic Development Bank

The Islamic Development Bank (IsDB) is a multilateral development institution, established in 1974 by member countries of the Islamic Conference Organization. There are currently 56 member countries, from Africa, Asia, and the Middle East.

IsDB's subscribed capital has increased substantially in recent years. It has launched a rapidly growing lending program and a number of initiatives aimed at specific objectives. As a result, IsDB has turned into a significant multilateral institution.

IsDB's authorized capital was kept small, at $2 billion, for about 18 years, from its establishment until the early 1990s. Since 1992, however, the authorized capital has been increased—to $6 billion in 1992, $15 billion in 2000, and $30 billion in 2006. IsDB's seven largest shareholders are Saudi Arabia, with 24.8%; Kuwait, with 12.3%; Libya, with 10.0%; Iran, with 8.7%; Egypt, with 8.6%; Turkey, with 7.8%; and the United Arab Emirates (UAE), with 7.0%. These shareholders occupy 7 of the 14 positions on the board of directors.

Four IsDB-associated entities have been established as separate multilateral institutions: Islamic Research and Training Institute (IRTI), Islamic Corporation for the Development of the Private Sector (ICD), the Islamic Corporation for the Insurance of Investment and Export Credit (ICIEC), and the International Islamic Trade Finance Corporation (ITFC). The *IsDB Group* refers to these four entities plus IsDB itself.

The IsDB Group has recently sharpened its vision to become the leader in fostering socioeconomic development in member countries and Muslim communities in nonmember countries. Its objectives include promotion of Islamic financial institutions, alleviation of poverty, and promotion of cooperation among member countries. Its areas of priority include infrastructure development, human development, agriculture, intra-trade (i.e., among member countries), private sector development, and research and development (R&D) in Islamic economics, finance, and banking.

IsDB utilizes various Sharia-compliant modes of financing to support its operations. It has a concessional vehicle for very poor countries, mostly in Sub-Saharan Africa and an ordinary vehicle for other countries. The engagement is legally structured to avoid interest bearing transactions that are prohibited in Islamic finance. IsDB's overall financial support has reached about $3 billion per year and has substantial room to grow; about 20 to 25% of the total goes to the energy sector. The power sector is the biggest recipient of these resources.

Among other IsDB Group institutions, the activities of ICD (established in 1999) are worth consideration in the financing of energy projects. ICD has an authorized capital of $1 billion, of which $282 million is paid in by its members. While designed after IFC, ICD incorporates Islamic finance principles. ICD has been in operation only for a few years and has already performed admirably, with an annual lending of about $200 million.

In addition to ICD, the IsDB Group in 2001 established the IsDB Infrastructure Fund (IIF), as a private investment vehicle to focus on infrastructure development in member countries. The objective of the IIF is to seek long-term capital appreciation by making equity and equity-related investments in infrastructure projects and infrastructure-related industries. The sectoral focus of the IIF is on power generation, petrochemical, and telecommunication, which comprise around 80% of its activities. The overall size of the IIF is $1.5 billion. Its offices are located in Bahrain, and it is managed by a private company—the Emerging Markets Partnership.

The OPEC Fund for International Development

The OPEC Fund for International Development is an autonomous multilateral development institution, established in 1976 by the 13 (now 12) members of the Organization of Petroleum Exporting Countries (OPEC). Its objectives are to promote cooperation among OPEC member countries and other developing countries by providing financial assistance in support of economic and social development.

Most of the OPEC Fund's assistance is provided in the form of project loans; however, some balance-of-payments support and structural adjustment programs are offered. Outright grants are extended in support of technical assistance, food aid, and research. Emergency relief aid is provided occasionally.

Geographical interests of the OPEC Fund include most countries in Africa and Asia and a few countries in Latin America, and energy is a sector of interest. The annual level of assistance is around $370 million, of which an average of 20%–25% is allocated to energy projects. The assistance includes loans and grants. In the case of loans, terms are decided on a case-by-case basis, depending on the project and the recipient country.

The Arab Fund for Economic and Social Development

The Arab Fund for Economic and Social Development is a regional development institution, established in 1972 through the initiative of the Arab League. Located in Kuwait, the Arab

Fund seeks to assist the 21 member countries of the Arab League in their efforts toward economic and social development. The fund's activities include financing development projects, participating in equity investments in private and public enterprises, and providing technical assistance grants.

The largest recipients of funds are Morocco, Egypt, Syria, and Tunisia. Moreover, most small African countries receive assistance. The average level of lending is $200–300 million per year. Energy projects account for 20%–30% of the total lending. The average lending terms are an interest rate of 4%, maturity at 23 years, and grace period of 5 years. Procurement is subject to international competitive bidding.

The Arab Bank for Economic Development in Africa (BADEA)

Established by the Arab League in 1973, BADEA is an autonomous regional development institution. Its member countries are Algeria, Bahrain, Egypt, Iraq, Jordan, Kuwait, Lebanon, Libya, Mauritania, Morocco, Oman, Qatar, Saudi Arabia, Sudan, Syria, Tunisia, and United Arab Emirates, as well as Palestine. Any Arab country is eligible to join BADEA. BADEA aims are to finance economic development in non-Arab African countries, to promote the movement of Arab capital, and to provide technical assistance for economic development. All member countries of the Organization of African Unity (OAU) that do not belong to the Arab League are eligible to apply for assistance. Almost every eligible African country has benefited from BADEA assistance. Energy is a sector of interest.

BADEA allocates between $160 and $200 million annually. The ceiling is 60% of the total estimated cost of the project, provided that it does not exceed $18 million. The interest rate ranges between 1% and 4%, with loan periods of 18–30 years and grace periods of 4–10 years.

6 Bilateral Sources Of Financing

Bilateral agencies are institutions set up and supported by their respective governments to provide investment and technical assistance to developing countries. The functions of bilateral agencies fall into two distinct categories:

- Provision of grants and highly concessional loans to developing countries on the basis of economic, social, and political considerations
- Provision of loans, guarantees, and insurance that are designed to advance the export of goods and services from the donor country and to promote the involvement of donor-country companies in projects in developing countries

In most donor countries, separate institutions are set up to handle the two functions. For example, in the United States, the U.S. Agency for International Development (USAID) is primarily responsible for development assistance, and the U.S. Export-Import Bank (USExim) promotes U.S. business abroad. However, in practice, the functions of the two agencies are intertwined. Furthermore, other public and private entities become involved, which results in overlap of activities within each donor country.

Development Assistance

All OECD countries have substantial development assistance programs. Official development assistance (ODA) comprises loans, grants, technical assistance, and contributions to multilateral institutions. Between 1990 and 2002, ODA increased by only $10 billion—from $70 billion to $80 billion. Since 2002, however, ODA has increased substantially owing to geopolitical reasons, such as combating terrorism and reconstruction of Iraq, Afghanistan, and Palestine. The OECD's development assistance reached a record high of $106 billion in 2005. In terms of volume, the United States and Japan are the largest donors. However, in terms of relative generosity, European countries make the largest contributions. Traditionally, Scandinavian countries, including Norway, Sweden, and Denmark, contribute at least 0.93% of their gross national income (GNI) to development assistance. In 2005, Luxembourg and the Netherlands gave around 0.83%, whereas Japan donated only 0.28%; although contribution from the United States has increased over the past five years, it is still only at 0.22% of its GNI. The figures for the United Kingdom, France, and Germany are 0.47%, 0.47%, and 0.36%, respectively.

The objectives of development assistance programs vary substantially. Assistance from the United States has a sharp focus on countries and regions that it considers to be important with respect to the maintenance of international stability. Japanese aid is aimed at strengthening economic relations. French assistance is based on strong historical ties. Many other European countries base their development assistance on social considerations.

In addition to OECD countries, developing nations such as Kuwait and Saudi Arabia provide assistance to poorer countries. These sources of assistance have become more important in recent years as increased oil revenue has enabled more generous contributions.

The themes of development assistance programs are somewhat different among donor countries. Most now consider poverty reduction, private sector development, and environmental preservation as high priorities.

Development assistance programs and agencies are normally parts of governments and are funded through public budgets. The funds are provided to developing countries in the form of grants or highly concessional loans. Some of these funds are channeled to developing countries through multilateral institutions—for example, the World Bank, regional development banks, and United Nations agencies. The rest of the funds are provided to developing countries on a bilateral basis.

Bilateral assistance programs provide funding to developing countries in support of

- Balance-of-payment requirements (i.e., the money will be at the disposal of the government of the recipient country to be disbursed against general imports)
- Specific investments (e.g., building a power plant or a pipeline)
- Preparation of a project (e.g., feasibility studies)
- More general technical assistance (e.g., strategic studies and training)

The last two items have become important components of bilateral aid programs. Most donor countries have established separate entities to handle the technical assistance program or have made it a major responsibility of the existing entities.

Export Financing

All OECD countries and some developing countries—such as Korea, India, and Brazil—provide concessional financing for export of goods and services by their own companies. The financing is provided either as a loan to the purchaser of the equipment (e.g., a power company in a developing country) or as a credit to the supplier (e.g., a U.S. manufacturer of power equipment). The loan can be made directly by the bilateral agency to either party or through a financial intermediary, such as a commercial bank. If the loan is made by a commercial bank, the bilateral agency insures repayment of the loan by the borrower.

Provision of export credit represents an important dimension in competition among the equipment suppliers of various countries. The presence and the extent of concessional finance sometimes distort price competition and may discourage efficiency of manufacturers. To avoid undesirable distortions, the governments of OECD countries in 1978 prepared the Arrangement on Guidelines for Officially Supported Credits, known less formally as the *OECD Consensus*. The arrangement sets limits on the favorability of terms of official export credit. All OECD countries agreed not to provide export credit exceeding 85% of the contract value and not to set interest rates below the OECD interest-rate matrix, which is revised semiannually. Although the arrangements have not been followed consistently, the exporting countries have further amended their rules through the following agreements:

- *Helsinki Package.* Agreed in 1991, this relates the level of tied aid to the gap in commercial viability of the corresponding projects.
- *Schaerer Package.* Agreed in 1994, this is related to two principle points: First, commercial interest reference rates (CIRRs) were introduced in place of subsidized interest rates. Second, a work program was agreed for the global untying of aid and analysis of export credits of agricultural products.
- *Ex Ante Guidance for Tied Aid.* Signed in 1996, this provides guidance for aid agencies to evaluate whether projects are eligible for tied aid under the Helsinki Package agreement. For instance, social projects are unlikely to be commercially attractive, hence could be considered eligible for tied aid.
- *Knaepen Package.* Agreed in 1997, this provides guiding principles for minimum risk-based premium fees for country risks.
- *Flexibility for Project Finance.* Agreed in 1999, this allows for greater flexibility in the timing and profile of repayments.
- *Common Approaches.* Agreed in 2001, this voluntary agreement has the aim of developing common procedures relating to the environmental impact of infrastructure projects supported by their governments' export credit agencies.

In the energy sector, agencies such as USExim and Japan Bank for International Cooperation (JBIC) play important roles in financing power and gas projects. Loans are normally made to the purchaser of equipment and services. Three methods are used to extend finance to an importing entity:

- *Direct lending,* where the bilateral agency signs a loan agreement with the importing entity. These loans are often tied to the purchase of equipment from manufacturers in the bilateral agency's country. The terms follow the OECD Consensus. There has been extensive criticism that tied loans in effect raise the cost of procurement and limit local ownership and trade liberalization. As a result, the relative portion of tied aid has declined systematically over the years. Some countries—for example the United Kingdom, Norway, and Ireland—have decided to make all their loans untied. Germany, Sweden, and Japan have made a large portion of their loans untied.
- *Lending through financial intermediaries* (also referred to as *relending* or *onlending*), where the bilateral agency provides the loan to a commercial bank, which relends to the importer.
- *Interest-rate equalization,* where a commercial bank provides a loan to the importer at a lower-than-commercial rate (normally the OECD Consensus rate). The commercial bank is then compensated by the corresponding bilateral agency for the difference between the commercial and the consensus rates.

The idea behind the second and third of these methods is to involve commercial banks in export financing. Many countries, most notably Japan, provide direct loans through negotiations between corresponding governments and within the framework of their overall aid programs to recipient countries.

Summary 6–1. Objectives of Bilateral Aid Agencies

- All countries of the Organisation for Economic Co-operation and Development (OECD) have set up agencies that provide financial and technical assistance to developing countries. These bilateral aid agencies provide grants and concessional loans to support general import requirements (balance of payment) or specific programs and projects.
- The support of bilateral aid agencies assists developing countries and sometimes promotes exports and other business interests of their own nationals.
- Most OECD countries have separate agencies for providing development assistance and export promotion. However, functions of these two agencies quite often overlap, and their funds may be combined to provide attractive financing packages.
- Export credit agencies often provide guarantees for loan repayment and insurance against political risks and other risks to commercial lenders. Some countries have separate agencies responsible for risk insurance.

INSURANCE AND GUARANTEES

In addition to providing support for developing countries, insurance and guarantees from bilateral agencies promote the sale of goods and services by the lending country's companies. Export credit insurance protects the exporter or the intermediary bank against commercial or political risks. Commercial risks include nonpayment by a nonsovereign or private sector buyer in the event of insolvency or failure to take shipments of goods; protection is also available against preshipment risks and unfair calling of the exporter's commitment bonds. Political risks include default by a sovereign entity, general moratorium on external debt, foreign exchange inconvertibility, cancellation of export-import licenses, delay in transfer of payments, war, and civil disturbances.

Insurance and guarantees from bilateral agencies (fig. 6–1) normally cover 85%–95% of losses caused by commercial risks and 100% of losses caused by political events. Coverage is either global or specific. Global coverage provides protection for all export transactions of a supplier or bank. Specific coverage provides protection for one transaction. Premiums for global coverage are lower than for specific coverage because of the spread of risks.

Summary 6–2. The Major Bilateral Development Assistance Agencies

- The largest contributors to bilateral development assistance are the United States, Japan, the United Kingdom, France, and Germany. Among other OECD countries, the Netherlands, Italy,

Canada, Sweden, Spain, Norway, Denmark, Belgium, Switzerland, Australia, and Austria make significant contributions.

- A few non-OECD nations, such as Kuwait, Saudi Arabia, Abu Dhabi, and South Korea, provide development assistance. The number of nations with export credit and support agencies is growing substantially. For instance, China, India, Brazil, Thailand, South Korea, Hungary, Uzbekistan, Israel, and Trinidad and Tobago all have such facilities.

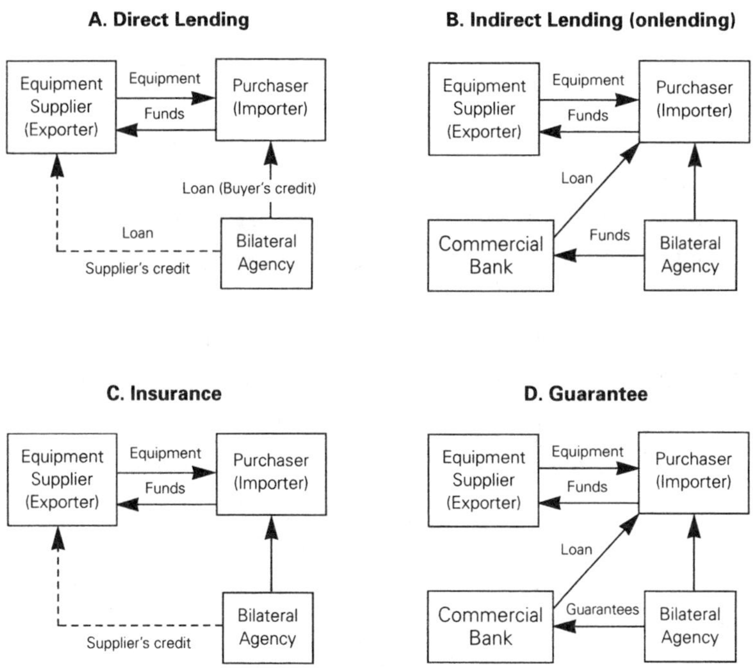

Fig. 6–1. Export credit insurance and guarantee facilities of bilateral agencies

United States

At present, the U.S. development assistance program contributes around $27.5 billion per annum in ODA. This has increased significantly from only $12.9 billion in 2002 mainly because of geopolitical reasons. In addition, the rise has also been due to the reconstruction of countries such as Afghanistan and Iraq.

The policy making and implementation of the development assistance program involve the following agencies:

- USAID is responsible for 50% of the U.S. aid in the form of grants to the recipient country.
- The Millennium Challenge Corporation (MCC) is a government-owned corporation responsible for the stewardship of the Millennium Challenge Account (MCA).
- USExim provides loans and loan guarantees for the procurement of U.S. goods and services. It has supported over $430 billion of U.S. exports worldwide.
- The Overseas Private Investment Corporation (OPIC) provides project financing through loans and loan guarantees, as well as political risk insurance and advisory services in support of U.S. investments and exports.
- The U.S. Trade and Development Agency (TDA) provides funding for large project-feasibility studies by U.S. firms.
- The U.S. Peace Corps provides grassroots assistance in approximately 70 developing countries.
- The U.S. State Department coordinates foreign policy aspects of U.S. assistance.
- The U.S. Treasury Department handles interactions with multilateral agencies.

In addition to providing grants to finance projects, USAID has numerous programs that support investment studies and preparatory work. Many of these programs provide assistance to the energy sector and have no direct link to promoting the interest of American companies; others support projects that do present potential business opportunities for U.S. companies. In the first category, studies supported by USAID have helped many developing countries to formulate energy development strategies, energy conservation and pricing policies, and so on. In the second category, several country-specific programs provide funds for work that promotes joint ventures of each specific country with U.S. companies. The Global Development Alliance program has led to USAID partnership with the private sector to introduce clean and efficient energy technologies. Similarly, a number of global institutes are working on the Methane to Markets Partnership. Other partnerships include the Global Village Energy Partnership, Efficient Energy for Sustainable Development, and Partnership for Clean Fuels and Vehicles.

The MCC was established in 2004 with a goal of reducing poverty by supporting sustainable, transformative economic growth in low-income countries. MCC field operations are at the earliest stages; nevertheless, the OECD considers the MCC to be a good example of how the U.S. aid system could be more broadly adapted in line with the aid-effectiveness principles of the Paris Declaration. This includes approaches to local ownership and funding that are results based (rather than tied only to U.S. procurement). Congress provided nearly $1 billion in initial funding and $1.4 billion and $1.7 billion in the following years for the MCC and its assistance programs.

The TDA was founded in 1981 and provides grants for feasibility studies and technical assistance, to promote U.S. exports to developing countries. The grant must be requested by a government entity in a developing country, but the work must be undertaken by a U.S. company. In reviewing grant requests, the TDA considers whether the study is likely to identify potential large-scale export opportunities for U.S. companies. After a grant request is approved, the TDA requests that the government of the host country issue the terms of reference to a short list of U.S. firms. The winner is selected through competitive bidding. Although the bidding process is managed by the relevant entity in the host country, the TDA pays the fees to the selected firm directly.

USExim operates as an independent U.S. government agency. Its president and chair is appointed by the president of the United States. The board of directors has five members (president, vice president, and three others), of whom only three can be from the same political party.

USExim's loans provide fixed-interest-rate financing for U.S. exports. USExim extends direct loans to foreign buyers of U.S. exports and intermediary loans to parties that extend loans to foreign buyers. USExim's rates are in line with OECD guidelines.

USExim guarantees provide repayment protection for private sector loans to creditworthy buyers of exported U.S. goods. USExim may also guarantee payments on cross-border or international leases. Most guarantees provide comprehensive coverage of both political and commercial risks, although they can be tailored to cover only political risks.

Summary 6–3. The U.S. Bilateral Assistance Program

- The United States is the largest provider of assistance to developing countries.

- The main agencies are the U.S. Agency for International Development (USAID), the Millennium Challenge Corporation (MCC), the U.S. Export-Import Bank (USExim), the Overseas Private Investment Corporation (OPIC), and the U.S. Trade and Development Agency (TDA).
- USAID has numerous support programs, many of which cover energy projects, studies, and other types of assistance requirements.
- The TDA provides grants for feasibility studies for proposed projects in developing countries.
- USExim provides loans and loan guarantees for procurement of U.S. goods and services.
- OPIC provides loans, guarantees, insurance, and equity to ventures involving significant capital and management participation by U.S. companies.

OPIC was spun off USAID in 1971. A U.S. government body reporting to the secretary of state, it provides loans, guarantees, insurance, and equity to ventures involving significant capital and management participation by U.S. companies. OPIC's objective is to support projects that improve U.S. global competitiveness, create American jobs, and increase U.S. exports. OPIC's loan and equity facilities are very small. Its guarantee and insurance authorization facility, however, is substantial (about $4 billion per year). Guarantees are issued for loans borrowed from U.S. financial institutions or foreign lending institutions that are at least 95% U.S. owned. OPIC's insurance covers political risks, including expropriation, political violence, and currency inconvertibility.

JAPAN

In the second half of the 1990s, Japan was the largest aid donor in the world, with a peak contribution of $15.3 billion in 1999. However, in 2001, aid dropped by $4 billion, mainly because of budgetary constraints, which resulted in Japan's being overtaken by the United States. Nonetheless, donations have recently increased back to more than $13 billion in an attempt to reach 0.7% of its GNI. In total, Japan has provided $200 billion over the past 30 years as part of its ODA program. Currently, 96% of aid is untied, and the ODA loan component is considerably greater than in other industrialized countries.

The program of Japanese assistance to developing countries provides financial and technical assistance to these countries while also recycling Japan's surplus funds. Allocation of funds in such a manner reflects an inherent ideology of peaceful coexistence with other countries, particularly within Asia. Hence, ODA provides a mechanism to actively support economic and social infrastructure development, human resource development, and institution building, as well as to establish reliable trade and investment links. The bilateral aid program includes an ODA component and an "other official flows" category, which refers to loans under the Capital Recycling Plan.

JBIC was established in 1999 by merging the activities of the Export-Import Bank of Japan (JEXIM) and the Overseas Economic Cooperation Fund (OECF). JBIC still operates under two accounts: the Overseas Economic Cooperation Operations (previously undertaken by OECF) and the International Financial Operations (previously undertaken by JEXIM). The former account is responsible for ODA processing and implementation, and the latter account is mainly responsible for the promotion of Japanese exports and imports.

Although JBIC is the main vehicle for Japanese assistance, several other entities are involved in planning, budgeting, and allocation of development funds. The Ministry of Foreign Affairs of Japan (MOFA) is officially the first point of entry and receives requests for grants and loans. Japan International Cooperation Agency (JICA) is currently in the process of merging with a loan department of JBIC and the grants administration department of MOFA (to be completed in 2008).

Developing countries in Asia receive about 85% of the Japanese aid, with top recipients being China, Indonesia, and the Philippines. Africa is also considered a priority region and receives about 8%–10% of the aid. In recent years, the proportion of Japanese ODA loans contributed to the electric power and gas sector has increased. The following examples are typical of JBIC assistance to the energy sector: In 2002, JBIC provided a loan of $200 million to a combined-cycle gas-fired power plant in Mexico, to support Japanese goods and services from the Mitsubishi Corporation. In 2005, JBIC lent $1.2 billion to support the Taweelah B power and desalination project in Abu Dhabi, as the equipment was supplied by Marubeni and JCC Corporation. Also in 2005, JBIC provided $1 billion of financing to Qatargas 3, where Mitsui gained a 1.5% equity stake in the facility and an off-take agreement.

Nippon Export and Investment Insurance (NEXI) is an independent governmental institution responsible for official export credit insurance. In 2001, it replaced the Export Insurance Division of the Ministry of International Trade and Industry (MITI). JBIC and NEXI work together to promote Japanese business. They both may provide support to a transaction, with JBIC extending finance and NEXI extending insurance against political risk—in particular, when other Japanese commercial banks are participating.

Summary 6–4. Japan's Bilateral Assistance Program

- Japan is the second-largest provider of assistance to developing countries.
- The main aid agencies are Japan Bank for International Cooperation (JBIC) and Japan International Cooperation Agency (JICA).
- Export support is provided through the International Financial Operations division of JBIC. Financial support for pre-investment studies, training, and other types of technical assistance is provided by JICA and JBIC.

Financial support for technical assistance—for example, pre-investment studies and training—is normally provided by JBIC. In addition, JICA is the official agency designated by the Japanese government to extend technical cooperation to developing countries. Its functions include funding sectoral and project-specific studies, providing expertise and equipment aimed at transfer of technology, and mobilizing finance from JBIC. Another agency operating in this area is the Japan Consulting Institute (JCI), which provides technical assistance for project preparation but does not provide funds for the completion of studies and preparatory work.

United Kingdom

The UK development assistance program has a long history that extends back to the 1930s, when the objective of the program was to assist the UK colonies. In more recent history, the responsibility for development assistance was with the Ministry of Overseas Development, which was established in 1974 and was subsequently converted in 1979 to a department under the Foreign and Commonwealth Office. In 1997, the Department for International Development (DFID) was created and is now headed by a cabinet-rank Minister of State. DFID's mission is to assist developing countries in poverty reduction, in partnership with governments, private sector, and civil society. DFID provides assistance to about 50 countries mostly in Africa and South Asia, as well as some low-income countries in other regions. Although DFID's emphasis is on social sectors, it does provide support to energy projects that have low-income characteristics, such as rural energy and energy supply to the poor.

Another important UK bilateral agency, Capital for Development (CDC), is a government-owned fund that invests in private equity of emerging markets in Asia, Africa, and Latin America. The Capital for Development Group was originally established in 1948 to strengthen the economies of the former UK colonies by providing finance for business. At the time when colonies gained independence, the entity was named Commonwealth Development Corporation (hence, the abbreviation CDC) and by 1970 was given authority to invest in poor countries beyond the Commonwealth. In 1997, the entity became a public-private partnership, renamed Capital for Development Group, and was given a mandate to focus exclusively on equity investments; the acronym CDC has been retained. CDC is called "a fund of funds," which means that it does not make direct investment in companies; instead, CDC works through an intermediated approach that is viewed suitable to leverage capital for the markets with great needs for investments. CDC has net assets of $2.8 billion. Energy projects, particularly power sector operations, represent a substantial part (37%) of the CDC's portfolio. Accessing CDC resources can help in the establishment of energy funds that are suitable vehicles for bringing private sector participation into energy projects.

Finally, the United Kingdom's official export credit agency is the Export Credits Guarantee Department (ECGD). Its objective is to help exporters of UK goods and services in their overseas business. Its instruments include guarantees, insurance, and reinsurance against various risks. Its guarantees include a buyer credit facility; supplier credit facility; lines of credit, which provide UK exporters of capital goods with a quick way to access finance made available by a UK bank to an overseas borrower; and guarantees to a lending bank. Its insurance facilities include an export insurance policy that protects an exporter against not receiving payments to which it is contractually entitled; a bond insurance policy, which provides protection in the event of a bond being called unfairly; and overseas investment insurance, which provides coverage related to political risks in developing countries.

France

France's overseas development assistance is channeled through the French Development Agency (l'Agence Française de Développement [AFD]). Its financial commitments reached €2.5 billion in 2005. Africa is the priority continent and receives more than half of the aid; the rest is allocated to countries in Asia and the Middle East.

AFD's overall mandate is poverty reduction. It provides support to various sectors including infrastructure and urban development. AFD emphasizes basic needs, such as energy access in poorer countries and opening competition in higher-income countries. It supports environmental protection operations, including energy efficiency and renewable energy. AFD often cofinances energy projects. Recent projects include electricity interconnection in Morocco, the Nam Theun 2 hydroelectric power project in Laos, and a natural gas project in Mozambique.

Policies governing export credit facilities are formulated by two government agencies: the Direction des Relations Economiques Extérieures (DREE) and the French Treasury. In practice, export support is provided through the Compagnie Française d'Assurance pour le Commerce Extérieur (COFACE), which is France's export credit agency. COFACE provides insurance against commercial and political risks. It also insures against political risk on direct investments in developing countries. COFACE is one of the world's most diversified export credit agencies, with its private operations being much greater than the French government's program.

Germany

The German development assistance program constitutes about $9 billion and has increased by about $2 billion since the 1990s. There are two distinct setups for development assistance and export promotion. The development assistance function involves the following agencies:

- The Federal Ministry for Economic Cooperation and Development (Bundesministerium für Wirtschaftliche Zusammenarbeit [BMZ]), which is responsible for policy making, decision making, and funding of the foreign assistance program. About 70%–80% of ODA comes from the BMZ budget and contributions from the Federal States.

- Kreditanstalt für Wiederaufbau (KfW), which handles the implementation of the BMZ's financial cooperation program. It is 80% owned by the German government and 20% by the Federal States. The KfW also provides mixed financing tied to procurement from German firms. Credits from the export-financing program are officially supported and are in line with OECD guidelines for official export credits.
- The Technical Assistance Corporation (Gesellschaft für Technische Zusammenarbeit [GTZ]), which is responsible for implementation of Germany's technical assistance program.
- Deutsche Investitions- und Entwicklungsgesellschaft, which is a member of the KfW Bankengruppe, promotes the private sector by providing loans and making equity investments. It also provides technical assistance to German firms making investments in the developing world.

The BMZ, as a policy-making body, decides whether a proposal fits into the German aid program. Subsequently, investment and export-financing proposals are passed on to the KfW for appraisal, and technical assistance proposals are sent to the GTZ. Both agencies handle implementation once the BMZ issues final approval.

In regard to export promotion, the Inter-Ministerial Committee is responsible for government policy. This committee is composed of representatives from the ministries of economics, finance, foreign affairs, and economic cooperation. Advised by members of the industrial, banking, export, and commercial sectors, the Inter-Ministerial Committee is chaired by the Ministry of Economics. The policies are then implemented by KfW, which can provide loans and guarantees.

In addition, the German government has selected Euler Hermes and Price Waterhouse to manage its three separate guaranty programs: export credit guarantees, investment guarantees, and untied-loan guarantees. These guarantees are issued on the account of the federal government. Export credit guarantees insure German exporters and banks against the nonpayment risk involved in export transactions; they cover a variety of risks, from construction to political risks. Investment guarantees support the investments of German companies in emerging economies. Untied-loan guarantees cover the loans awarded by German lenders to borrowers in emerging countries; this type of guarantee is aimed not at export promotion but instead at other interests of German government, such as securing energy resources.

Netherlands

The Netherlands development assistance program involves around $5.1 billion per year and is implemented by the Directorate General for International Cooperation (DGIS) within the Ministry of Foreign Affairs. The Netherlands is one of the most generous donors, providing around 0.82% of its GNI. The program strongly focuses on the eradication of poverty.

The development assistance program has a special focus on energy, administered by the Climate, Energy, and Environmental Technology Division. Support has been directed toward renewable energies, energy efficiency, and rural energy in Africa, Asia, and Latin America.

Insurance against export credit risks is provided by the Nederlandsche Credietverzkering Maatschappij (Atradius), a private company that receives support from the government.

Italy

Italy's development assistance program involves about $5 billion and is implemented by the Directorate General for Development Cooperation (DGCS) of the Ministry of Foreign Affairs. Although a major part of past assistance has been in the gas and power sectors, the authorities have now shifted their emphasis to social services and governance issues, resulting in lower allocations to the energy sector.

The geographical distribution of Italy's aid is concentrated in poorer regions. Nearly 50% is channeled to sub-Saharan Africa.

Insurance against political and commercial risks is provided for export credits by the Sezione Speciale per l'Assicurazione del Credito all'Esportazione (SACE), which is quite active in the energy sector.

Canada

Canada's development assistance program constitutes around $3.7 billion per year and is likely to increase in the coming years, as Canada is committed to increasing development aid to around 0.7% of its GNI. The funds are channeled through the Canadian International Development Agency (CIDA), which is an autonomous agency under the Foreign Affairs and International Trade Department. About 25% of the ODA goes to basic human needs, with a special focus on Africa. However, CIDA has several bilateral programs focused on energy efficiency and clean energy. In recent years, it has supported the Tata Energy Resources Institute's energy-efficiency project in India and has initiated the Canada-

China clean-coal program. In addition, the Canada Climate Development Fund is focused on reducing carbon emissions and can be tapped for the financing of energy projects.

Export Development Corporation (EDC) provides export credit with an interest in oil, gas, and power. EDC provides both insurance and financing to help Canadian exporters compete in other countries.

Sweden

The Swedish development assistance program has been praised because of its generosity and its emphasis on poverty reduction and the environment. Sweden contributes $3.2 billion per annum—and nearly 1% of its GNI—to development assistance. Sweden bases its development assistance program on the philosophy of helping the underprivileged. The main aid agencies are as follows:

- The Ministry of Foreign Affairs and the Export Credit Guarantee Board, which are responsible for setting policies
- The Swedish International Development Cooperation Agency (SIDA), which implements the ODA policies, including grants and export credits

Norway

Coordination and implementation of the Norwegian development assistance program was primarily the responsibility of the Norwegian Agency for Development Cooperation (NORAD). However, since 2004, the role of NORAD has been diminished to that of an advisor to the government of Norway. Instead, the responsibility for all development aid has been concentrated in the Ministry of Foreign Affairs. At the same time, the objective of the assistance has also been focused more on poverty reduction. Still, Norway is a strong supporter of clean-energy initiatives and provides assistance to projects that aim to improve energy efficiency or increase energy access to the poor.

Establishment of the Norwegian Investment Fund for Developing Countries (Norfund) in 1997 added a new instrument of Norwegian bilateral assistance to developing countries. Norfund's mandate is to provide loans and risk capital to viable private ventures in developing countries.

Denmark

Denmark's annual assistance constitutes about $2.1 billion per year. Denmark is one of the most generous donors in the Western world, contributing nearly 1% of its GNI. Sectors of concentration include agriculture, social and economic infrastructure, industry, and, on a relatively small basis, energy.

The Industrialization Fund for Developing Countries (IFU) was established in 1967 to support joint ventures between Danish companies and enterprises in developing countries. It has supported several small investments in oil, gas, and power in Latin America.

Export support is provided by the Export Credit Council (Eksportkreditraadet [EKR]) and the Danish Export Finance Corporation (EFC). The EKR offers only guarantees, whereas the EFC provides loans with the condition that the EKR provide guarantees. The EKR's guarantee fees depend on the type of coverage (for commercial versus political risks), as well as the creditworthiness of the country and the borrowing entity. Interest in the energy sector, for both agencies, is limited to efficiency improvement and promotion of alternative energy. The EFC has in particular supported wind energy in a number of countries, including China, Egypt, India, and the Philippines. Denmark is a market leader in the manufacturing of wind turbines, which has triggered financial support in this area.

Switzerland

The Swiss development assistance program consists of about $1.8 billion, amounting to around 0.44% of its GNI. Swiss aid is managed by two agencies: the Swiss Agency for Development and Cooperation (SDC) and the State Secretariat for Economic Affairs (SECO). Both are part of the Federal Department of Economic Affairs (DEA). The primary objective of Switzerland's development assistance policy is to alleviate poverty.

The Export Risk Guarantee Agency (ERG) is the official credit agency in Switzerland. However, concessional funds are also made available for export credit by the Federal Office of Foreign Economic Affairs.

Kuwait

The Kuwait Fund for Arab and Economic Development (KFAED) was established in 1961 as a mechanism through which the State of Kuwait could extend loans and aid to other Arab countries. The mandate was modified in 1974 so that KFAED could provide assistance to other (non-Arab) developing countries as well. KFAED is historically significant because it was the first of its kind. It is also financially significant owing to the size of its capital (more than $6 billion).

KFAED is the channel for all of Kuwait's development assistance, including contributions to multilateral development institutions and other specialized funds. On its own, KFAED provides direct assistance to developing countries through loans, guarantees, and grants for technical assistance. Its sectors of emphasis include agriculture and infrastructure, including the energy sector. Loans and grants to the energy sector account for about 35% of KFAED's development assistance program. The regional composition of its portfolio is still concentrated in Arab countries but is gradually expanding to include other countries in Africa, Asia, Latin America, and Central and Eastern Europe.

Saudi Arabia

The Saudi Fund for Development (SFD) was established in 1975, with an initial capital of about $3 billion. The capital was increased to about $10 billion in 1991. SFD has contributed to the financing of development projects in 15 Arab, 30 African, 13 Asian, 5 European, and 5 Latin American countries. The terms of the loans are rather generous: the interest rate is about 1%, and the repayment period may be as many as 50 years.

SFD was initially limited to the financing of projects in developing countries. The mandate was amended in 1999 to enable SFD to provide credit and insurance in support of national non-oil exports.

SFD's development finance activities do not have any geographical or sectoral restrictions but rather assign priority to lesser developed or low-income countries. SFD's exposure is limited by covering up to 50% of the project cost and by keeping its contribution to each project and each country below 5% and 10%, respectively, of the fund's capital at any one time.

SFD works directly with the government or public entities of each country. A major part (about 20%) of its program in the past two decades has involved working with energy utilities, particularly electricity transmission and distribution companies.

Korea

Korea was a major recipient of development aid for several decades—as recently as the 1990s, when it faced a serious financial crisis. The government takes pride in its successful use of development assistance and has attributed many of its economic accomplishments to such aid. It has, therefore, a genuine desire to provide assistance to developing countries to finance their investment and institutional capacity building.

Korea's intent to help other countries was reflected in a program introduced in the 1980s for the purpose of sharing its experience of rapid economic growth with other developing nations. The program was formalized in 1989 by the establishment of the Economic Development Cooperation Fund (EDCF), through which concessional loans for development projects were provided to the governments of developing countries. In addition, in 1991 the Korea International Cooperation Agency (KOICA) was established to manage grant aid and technical cooperation programs.

Presently, Korea's bilateral assistance is channeled through two distinct mechanisms. Loans are provided by EDCF and implemented by the Export-Import Bank of Korea (KExim). Grants and technical cooperation programs are implemented by KOICA under the policy guidance of the Ministry of Foreign Affairs.

The overall volume of Korean ODA is still small, around $450 million per year. However, the level of support is growing rapidly. The regional focus of the assistance remains Asia; China is the largest recipient of Korean assistance. However, there is an emerging tendency for extending assistance to other regions. Energy sector activities include rehabilitation of power projects and electricity networks.

KExim has also taken an aggressive approach to supporting Korea's export of goods and services. Its activities supplement those of the Korea Export Insurance Cooperation (KEIC). KExim's program has expanded rapidly over the past decade, reaching about $25 billion in 2005. Export credit accounts for 80% of this volume, and overseas investment credit accounts for another 9%. Of the overall loans, about 46% were disbursed in Asian countries, and 14% were disbursed in the Middle East.

China

China's well-recognized ability to implement major construction projects has made it a desirable development partner in many parts of the world, particularly the Middle East. In recent years, it has added to its comparative advantage the possibility of financing major infrastructure projects.

The responsibility for concessional finance is given to the Export-Import Bank of China (China Exim Bank), which was established in 1994. The mission of the China Exim Bank is to promote the export of Chinese goods and services and to strengthen relations with foreign countries. Its products include various banking functions, as well as export credit and concessionary loans.

Concessional loans are extended by the bank on behalf of the government of China, to promote economic development in the recipient counties and to boost economic cooperation between these countries and China. These loans are awarded to projects that are approved by the government of China and the government of the recipient country, with the conditions that Chinese companies be selected as equipment manufacturers or contractors and that no less than 50% of the procurement come from China. Although the concessional loans promote Chinese exports, the programs for buyer's credit and seller's credit aim more directly at export promotion. However, under these latter programs, the financial terms are more commercial.

China Exim Bank is particularly active in the energy sector. A few recent examples include financing of the China National Petroleum Corporation's (CNPC) acquisition of an oil field in Sudan and a $3 billion program in Angola to develop infrastructure projects in return for supplying 10,000 barrels per day (b/d) of oil to China. Financial assistance to these African countries has paved the way for the acquisition of significant oil concessions. Aid has also created significant opportunities for Chinese companies. In 2006, there were 900 projects in Africa that involved Chinese firms.

TABLE 6–1. MAJOR AGENCIES INVOLVED IN BILATERAL FINANCING

Major Agencies Involved in Bilateral Financing

Country	ODA 2005 ($ million)	ODA/GNI, 2005	Development Assistance Agency	Export Credit Agency	Insurance and Guarantee
United States	27,622	0.2	USAID	Export-Import Bank of the United States (Ex-Im Bank)	Export-Import Bank of the United States (Ex-Im Bank)
			MCC (Millennium Challenge Corporation)	Private Export Funding Corporation (PEFCO)	Overseas Private Investment Corporation (OPIC)
				United States Trade and Development Agency (USTDA)	
Canada	3,756	0.34	Canadian International Development Agency (CIDA)	Export Development Canada (EDC)	Export Development Canada (EDC)
Japan	13,147	0.28	Japan International Cooperation Agency (JICA) Japan Bank for International Cooperation (JBIC)	Japan Bank for International Cooperation (JBIC)	Nippon Export and Investment Insurance (NEXI)
United Kingdom	10,767	0.47	Department for International Development (DFID)	Export Credits Guarantee Department (EGGD)	Export Credits Guarantee Department (EGGD)
			Capital for Development (CDC Group)	UK Trade and Investment (UKTI)	
Germany	10,082	0.36	Federal Ministry for Economic Cooperation and Development (BMZ)	KfW Development Bank	Foreign Trade and Investment Promotion Scheme (AGA)
			Agency for Technical Co-operation (GTZ)		
			KfW Development Bank		
France	10,026	0.47	French Development Agency (AFD)		COFACE
Italy	5,091	0.2	Ministry of Foreign Affairs (DGCS)	SACE S.p.A. (Servizi Assicurativi del Commercio Estero)	SACE S.p.A. (Servizi Assicurativi del Commercio Estero)
Netherlands	5,115	0.8	Ministry of Foreign Affairs		
Sweden	3,362	0.94	Swedish Intrnational Development Cooperation Agency (Sida)	Swedish Export Credit Corporation	Export Credit Guarantee Board (EKN)
Norway	2,786	0.9	Ministry of Foreign Affairs	Eksport finans	Norwegian Guarantee Institute for Export Credits (GIEK)
			Norwegian Agency for Development Cooperation (Norad)		
Denmark	2,109	0.8	Ministry of Foreign Affairs	Export Finance Corporation	Eksport Kredit Fonden (EKF)
Belgium	1,963	0.53	Directorate-General for Development Cooperation (DGDC)	Belgian Corporation for International Investment (BMI-SBI)	Belgian Export Credit Agency (ONDD)
			Belgian Technical Cooperation (BTC)		
Finland	902	0.4	Ministry for Foreign Affairs	Finnvera	Finnvera

Spain	3,018	0.27	Secretaría de Estado de Cooperación Internacional (SECI)	Instituto Español de Comercio Exterior (ICEX)	Compañía Española de Crédito a la Exportación (CESCE)
			Agencia Española de Cooperación Internacional (AECI)		
Switzerland	1,767	0.44	SDC - Swiss Agency for Development and Cooperation		
			State Secretariat for Ecomomic Affairs (SECO)		Swiss Export Risk Insurance (SERV)
Austria	1,573	0.5	Austrian Development Agency (ADA)	Oesterreichische Kontrollbank Aktiengesellschaft (OeKB)	Oesterreichische Kontrollbank Aktiengesellschaft (OeKB)
Australia	1,680	0.2	AusAid	Export Finance and Insurance Corporation (EFIC)	Export Finance and Insurance Corporation (EFIC)
Ireland	719	0.4	Irish Aid (Department of Foreign Affairs)		The Insurance Corporation of Ireland (ICI)
Turkey	601	0.1	Ministry of Foreign Affairs	Export Credit Bank of Turkey (Turk Eximbank)	Export Credit Bank of Turkey (Turk Eximbank)
Greece	384	0.1	YDAS (Hellenic Aid) under the MOFA		Export Credit Insurance Organization (ECIO)
Portugal	377	0.21	Portuguese Institute for Development Support (IPAD)		Companhia de Seguro de Creditos, S.A. (COSEC)
New Zealand	274	0.27	New Zealand International Aid and Development Agency (NZAID)	New Zealand Export Credit Office (NZECO)	New Zealand Export Credit Office (NZECO)
Luxembourg	256	0.84	Lux-Development (Luxembourg Agency for Development Cooperation)	L'Office du Ducroir	L'Office du Ducroire
Poland	205	0.07	Polish Aid (under the MOFA)		Export Credit Insurance Corporation
Czech Republic	135	0.1	Ministry of Foreign Affairs	Czech Export Bank	Export Guarantee and Insurance Corporation (EGAP)
Hungary	100	0.1	Ministry of Foreign Affairs	Hungary Export Import Bank	MEHIB (Hungarian Export Credit Insurance)
Slovak Republic	56	0.12	Agency for International Development Cooperation – Slovak Aid (under MOFA)	EXIMBANKA SR	EXIMBANKA SR
Iceland	27	0.18	Icelandic International Development Agency (ICEIDA)		
Korea	752	0.1	Korea International Cooperation Agency (KOICA)	Korea Eximbank	Korea Export Insurance Corporation (KEIC)
			Economic Development Cooperation Fund (EDCF) - managed by Eximbank		
Saudi Arabia	21		Saudi Fund for Development (SFD)	Saudi Export Program (SEP)	
Kuwait			Kuwait Fund		
China			China Exim Bank	China Exim Bank	China Exim Bank

7 Accessing Commercial Funds

Historically, projects and companies located in developing countries have had less access to commercial sources of finance as compared with those in the industrial world. However, the situation is changing rapidly with the introduction of new instruments. Moreover, significant changes are taking place in the financial markets of countries such as China, India, and Russia. Another emerging change is that both private and public (state-owned) energy companies are now accessing commercial sources of funds. Indeed, in some developing countries—for example, China, Algeria, and Brazil—a significant portion of the investment needs of public utilities is covered through commercial sources of funds.

Commercial sources of funds can be classified in several ways. One convenient way is to divide them broadly into domestic sources and international sources, each of which could contribute to debt or equity. *Specialized energy or infrastructure funds,* which are often a source of combined domestic and international funds, have recently been developed as a further vehicle of finance.

Commercial funds for energy companies in developing countries come from six major sources:

- Domestic capital markets, which provide both equity and debt financing
- International commercial banks, from which companies can get independent loans from several banks or syndicated loans (provided by a number of banks on a pro rata basis under identical terms and conditions)
- International equity markets, where a company's equity shares can be sold through public offerings or private placement with institutional investors (pension funds, insurance companies, mutual funds, etc.)
- International bond markets, where a company can borrow funds by issuing bonds through public offering or private placement with commercial banks and institutional investors
- Specialized energy funds, which are sometimes created with partial government sponsorship, to facilitate private sector investment in the energy sector, and other times created purely on a commercial basis, to channel funds from investors in the world's major financial centers to projects in developing countries
- Financial contributions from project beneficiaries, including equipment suppliers and contractors (who may provide equity or debt finance to win the business of building the project facilities), sellers of fuel to the project plant, and purchasers of project output (who may make financial contributions because they want to do business with the project company)

Summary 7–1. Commercial Sources of Finance

The major sources of commercial finance for energy projects in developing countries are as follows:

- Domestic funds, including commercial banks, bond markets, and equity markets
- International commercial banks, which can either provide stand-alone loans or participate in syndicated loans
- International bond markets of the United States, Japan, Europe, and the United Kingdom
- International equity markets
- Specialized energy or infrastructure funds
- Equity and debt finance contributions from project beneficiaries

Domestic Funds

An energy company can mobilize debt or equity from domestic sources. In some countries, commercial banks are the main vehicle of debt finance. Because of the large size of energy projects, domestic commercial banks provide syndicated loans—that is, a number of banks join together in a coordinated fashion. In quite a few countries, the domestic bond market has become a major source of commercial borrowing by energy companies—public utilities in particular.

Mobilizing equity for energy projects from domestic commercial sources was not traditionally considered to be feasible, because the equity markets could not provide such large sums. This trend is changing as the equity markets of some countries have grown in size and liquidity.

It is unrealistic to expect availability of domestic funds for energy projects in all small and poor economies. However, for most developing countries, the use of domestic sources of funds is proving to be feasible. Recent trends indicate that most such countries are moving toward a situation in which a combination of equity markets, commercial banks, and the bond market provide a significant amount of funds for energy projects. Nevertheless, at present, the situation varies from one country to another. In some countries, such as Russia, the bond market has become the dominant source of funds; in other countries, such as China, commercial banks represent the major source; and in others still, such as some Middle East countries, the equity markets are tapped at large scales to fund energy projects.

In the developing countries of the Asia region, syndicated loans, equity markets, and debt markets have become viable sources of financing for energy projects. Since the Asian crisis of 1990s, the authorities have focused on the strengthening of financial institutions. Multilateral institutions such as the ADB have assisted the countries in establishing robust and flexible regulations. Authorities have successfully promoted the bond market by establishing rating agencies and by relaxing some of the regulatory requirements. The significant presence and growth of institutional investors also means that private placement is an attractive avenue to raise funds for energy. Assets under management of insurance companies have grown in Singapore, Malaysia, and Korea from 19%, 14%, and 14%

in 1998 to 36%, 19%, and 25% of GDP, respectively, by 2004. This growth, along with a host of other factors, has stimulated the growth of the corporate bond market in Malaysia and Korea, which represented only 8% and 15% of GDP in the early 1990s and grew to an average 43% and 26% of GDP, respectively, in 2004.

In China, the domestic financing of energy projects has grown substantially but comes primarily from the banking sector. Lending has increased from 86% of GDP in 1990 to 149% of GDP by 2005. Even though the governmental bond market is relatively developed, the corporate bond market is still nascent. Corporate bonds make up only 3.5% of all bonds, with Ministry of Finance T-bonds and central bank bills making up 46% and 24%, respectively. Nonetheless, prospects for raising finance through the issuing of bonds are improving. In 2004, there was some relaxation of approval limits on corporate bond issues, and regulatory changes allowed easier access of nongovernmental firms to the bond market. Furthermore, in an attempt to stimulate the market, in 2005, the IFC issued an Rmb1.13 billion ($138 million) 3.4% 2015 bond. This was the first domestic bond to be issued by a foreign issuer.

The Chinese stock markets are a relatively new development: the Shanghai and Shenzhen exchanges both opened in 1990. The capitalization is about $500 billion, compared to over $4.5 trillion in Japan and $14 trillion in the United States. However, a new surge of initial public offerings (IPO) has changed the prospects for funding corporate investments.

In India, the domestic market has been the principal avenue for the funding of energy projects. Poor regulation has limited the development of the bond market, but the equity market has been a major vehicle for finance. There is no restriction on foreign investment in the equity market. Consequently, established firms have gone to the equity market to fund their new investments. For instance, in 2005, India's Reliance Petroleum raised $600 million through an IPO on the Mumbai stock exchange and another $600 million through private placement of its shares with institutional investors to help finance a 580,000 b/d export refinery in Jamnagar.

In Russia, the bond market is an important source of financing for energy projects. Since the 1998 financial crisis, bank loans have not been a major vehicle of finance for energy project companies. The corporate bond market, by contrast, has developed into a very important avenue for finance. The growth of the market was largely the result of limitations on other avenues of funding. Since the first corporate bond was issued in 1999, the total value of outstanding corporate bonds has increased from 39 billion rubles to 267 billion rubles in 2005. The Russian corporate bond market is particularly attractive owing to the significant participation of foreign investors.

The Russian stock market has not yet become a significant source of finance for the energy sector. Energy projects have found that raising funds by the issuing of shares domestically is not desirable. This is because of high costs of listing, low valuations, and high volatility. Hence, whenever possible (especially as a result of the high profile of the energy sector), Russian firms have preferred the international markets.

Perhaps the most unexpected change that has occurred in recent years is in the way that some countries in the Middle East, particularly those of the Gulf Cooperation Council (GCC), are accessing domestic funds. This change has been triggered in part by the abundance of oil money and in part by innovative measures that were taken to cope with the requirements of the Islamic laws.

The GCC countries have historically been a major focus of energy investments. In recent years, the possibility of raising funds domestically has increased, particularly in Dubai, Qatar, Bahrain, and Saudi Arabia. The equity markets have been an important avenue for funds. In the past decade, a significant portion of funds from oil revenues have been injected into the domestic and regional markets, resulting in a booming stock market. This in turn has enabled firms to raise equity from domestic sources. In 2005, the Yanbu petrochemical complex raised about $500 million through an IPO for a project that was at the concept stage. The IPO was oversubscribed by about 200%.

The GCC bond markets did not develop in the past partly because of the limitations of Islamic banking. However, there has been significant progress in recent years, as companies have successfully tapped the GCC bond markets by using Sharia-compliant instruments. IsDB launched the first *sukuk* (Islamic bond) in 2001. Since then, the volume has expanded exponentially.

Islamic finance has also become a niche market for many international and Western financial institutions, which have established specialized divisions to develop Sharia-compliant products. As Islam prohibits the payment or the receipt of *riba* (usury/interest), bonds need to be structured to pay a return or a share of profit, rather than interest. The following are the most commonly adopted structures of sukuks in project finance:

- *Ijara*: Bonds are backed by an asset under a leasing contract. An Islamic bank leases the asset to a client for rent. The rent is fixed so that the bank gets profit in addition to repayment of the original investment.
- *Istisna'a:* This is a manufacturing contract. A firm agrees to make certain goods (services) with predetermined specifications at a given price and for a fixed date of delivery. For instance, the Dolphin Gas Project, which includes a pipeline from Qatar to the UAE, obtained $1 billion over four years for bridge financing through an ijara and an istisna'a; the ijara principle was adopted for the sale leaseback of operational assets, and the istisna'a principle was used as a forward lease of assets not yet in service.
- *Murabaha:* Bonds are based on a cost-plus sale contract. The seller purchases an asset on behalf of the buyer and passes it on to the buyer while adding some profit. Hence, it is not a loan that charges interest, but rather a sale of an asset for a deferred price. For instance, Saudi Arabia Basic Industries Corporation (SABIC) has financed much of its domestic expansions with this type of sukuk.
- *Musharaka:* A profit-and-loss–sharing contract. All the partners will contribute to the financing of the project and will then share profits based on a predetermined ratio and losses according to each party's equity stake.

Finally, a growing number of specialized energy or infrastructure funds are being established in developing countries to fund energy projects. As explained in detail later in this chapter, energy funds were initially established to transfer resources from investors in industrial countries to projects in developing countries. However, in recent years, new ventures have emerged that are country specific. They are established with seed money from the sponsors but try to get most of their additional funding from the equity market through an IPO or the debt market, including bonds, private placement, and bank borrowing.

Summary 7–2. Domestic Funds

- Energy projects can raise equity from domestic sources, including commercial banks, bond markets, and equity markets.
- The composition of domestic funds varies depending on the country's rules, regulations, and investment climate. Bank loans make up the bulk of financing in China and India, while the bond market represents the major source of funds in Russia. In India and parts of the Middle East, the equity market has become a major source of funding for energy projects.
- Specialized energy (or infrastructure) funds are also used as a vehicle to channel money from the stock market or the bond market to energy projects.

International Commercial Sources

The traditional sources of international finance include international commercial banks, bond markets, and equity markets. However, there are now a variety of instruments and arrangements that offer greater potential for accessing international sources. The choice of one such instrument or any combination thereof would depend, among other factors, on the nature of the project, the credibility of the sponsors, and the country in which the project is located.

The role that international commercial banks play in the funding of energy projects has changed significantly in recent years. Up to the late 1990s, commercial banks provided loans to corporations investing in the energy sector. Often these were large-scale loans, because most energy projects needed sizeable financing. The interest of international commercial banks in developing countries dropped sharply in the 1980s, when certain loans to Latin American countries went into default. Interest reemerged in the 1990s, as countries in that region and in East Asia offered significant business opportunities. Unfortunately, this was also followed by the East Asia and Russian crises of the late 1990s, which again resulted in bad debt. Therefore, international commercial banks once again became cautious about exposing themselves to projects in developing countries—in particular, those in the energy sector, which were significantly affected during the financial crises.

Today, international commercial banks are very active in energy projects in developing countries for several reasons. First, the energy business has grown substantially owing to high oil prices since the early 2000s, resulting in the formulation of numerous energy projects. Second, international commercial banks have changed the manner of their participation from stand-alone to syndicated loans; this has allowed them to limit their exposure while participating in large transactions. Third, commercial banks have adopted the role of investment banking, which is focused on raising funds (in both debt and equity) from all possible sources, often through underwriting and financial advisory services.

In the traditional sense, investment banks differ from commercial banks that are in the business of taking deposits and making loans. Indeed, in the United States, the Glass-Steagall Act prohibited banks from simultaneously accepting deposits and underwriting securities. This law was repealed in 1999, providing the flexibility for engagement. In recent years, the lines between investment and commercial banking have blurred, especially as commercial banks have offered more investment-banking services. Today, there are still firms that are specialized in investment banking, but most large players have turned into financial conglomerates that combine commercial and investment banking and sometimes insurance—combinations of business that were illegal in the United States prior to 1999.

By playing multiple roles and offering a variety of instruments, international banks have become major players in the funding of energy projects in developing countries, particularly when debt is raised on a nonrecourse or a limited-recourse basis. Commercial banks participate in loan syndications but also underwrite securities—that is, they raise money on behalf of corporations that are issuing (both debt and equity) instruments. In the case of syndicated loans, one or several of the participating firms may play the role of *mandated lead arranger* (MLA), which structures the entire loan, takes part of the loan, and syndicates the rest to other banks. Participating firms may also play the role of *book runners,* which manage or lead the process of underwriting and maintain the books of securities sold for an issue.

TABLE 7–1. PROMINENT COMMERCIAL FINANCIAL INSTITUTIONS ACTIVE IN THE ENERGY SECTOR

Banking institution	Country	Total Assets (31 Dec 2005, $ million)
ABN Amro	Netherlands	1,039,088
Bank of America	USA	1,291,803
BNP Paribas	France	1,484,161
Calyon	France	567,811
Citigroup	USA	1,494,037
Credit Suisse Group	Switzerland	1,018,832
Deutsche Bank	Germany	1,170,415
Fortis Group	Belgium/Netherlands	859,998
Goldman Sachs	USA	706,804*
HSBC Bank plc	UK	665,384
ING Bank NV	Netherlands	983,915
JPMorgan Chase & Co.	USA	1,198,942
Mizuho Bank	Japan	643,858
Rabobank	Netherlands	597,207
Banco Santander Central Hispano	Spain	954,507
Societe General (SG)	France	1,022,189
Sumimoto Mitsui Banking Corp.	Japan	859
UBS AG	Switzerland	1,568
Mitsubishi UFJ	Japan	171

* Data as of 11/24/2005

Source: Fitch Ratings Ltd (http://www.fitchresearch.com)

Table 7–1 names the commercial/investment banks that are most active in the energy sector of the developing world. Chapter 10 describes in detail how these banks have participated in major energy projects. For example, in the case of the Tangguh liquefied natural gas (LNG) project in Indonesia, a debt issue of about $1 billion was signed with seven banks: the Bank of Tokyo-Mitsubishi, BNP Paribas, Fortis Bank, ING, Mizuho, Standard Chartered, and Sumitomo Mitsui Banking Group. The financial advisor for this transaction was Societe Generale.

Increasingly, international commercial banks join forces with domestic banks in syndicating loans and underwriting securities. In a major petrochemical project in Saudi Arabia, a combination of 20 banks participated in a $500 million loan syndication. About half of these banks were domestic or regional banks such as Gulf International Bank, Arab Banking Corporation, the Saudi British Bank, and the National Commercial Bank. The international commercial banks included such well-known entities as ABN Amro, BNP Paribas, Citibank, Fortis Bank, ING, and UFJ Ltd.; ABN Amro played the roles of financial advisor and underwriter of the entire debt.

Summary 7–3. The Role of International Commercial Banks

- The role of international commercial banks has changed significantly in recent years to combined commercial and investment banking.
- International banks not only participate in syndicated loans but also underwrite debt and equity issues and provide financial advisory services.

- The aforementioned business combination, along with significant growth in energy business, has resulted in sharply increased participation of commercial banks in developing countries.

International Bond Markets

The major bond markets are the United States, Japan, the euro area, and the United Kingdom. These markets are tapped primarily by domestic borrowers, including federal and local governments and private corporations. However, with globalization and the growth of rating agencies, bonds issued by foreign firms have expanded substantially.

The U.S. bond market provides the greatest access to funds. It is the largest securities market in the world, around $14 trillion in size. New issuances accounted for $10 trillion in 2006, with a turnover of $500 billion each day. The U.S. bond market represents around 47% of the developed world in terms of outstanding volume. Government debt makes up a substantial proportion; however, it is on decline. U.S. treasuries accounted for over 31% of debt in 1999. This has decreased to less than 16% currently. By contrast, corporate bonds have been growing rapidly and now equal 22% of the U.S. GDP. Foreign bonds have also grown in the past 20 years. This market, however, is dominated by highly rated sovereign and sovereign-guaranteed issuers.

The Japanese bond market is also dominated by government securities and financial institutions, with outstanding debt at 141% and 25% of GDP, respectively. However, the corporate bond market is the second largest in the world and is growing. Outstanding corporate debt has grown from around 2.6% of GDP in the early 1990s to over 16% of GDP in 2006. The catalysts were the abolition of issue standards in 1996 and the financial crisis, which restricted bank lending. The corporate bond market is mostly bought by Japanese institutional investors.

The euro-area bond market is growing rapidly as the EU is expanding. The increasing importance of the euro as a currency has led to growth in euro-dominated issues. The advantage of issuing bonds in euros is that it provides a means to diversify sources of funding away from the U.S. dollar. It also provides access to a large market with a very high average savings rate. In 2001, 36.5% of all new issues were in euros, compared to 49.1% in dollars. In 2006, the issuance of euro-denominated bonds was at the same level as that of dollar-denominated bonds. Although the euro-area corporate bond market is smaller than that of the United States and Japan, it is growing rapidly. The share of the market in world total has grown from 8% to 19% since the introduction of the euro as a common currency.

The U.K. bond market has been an important finance avenue for foreign companies, but it is now becoming less significant owing to the growth in the euro area. The U.K. bond market is still dominated by government debt. Of the four major bond markets, it is the only one that has demonstrated a decline in outstanding domestic debt as a percentage of GDP over the past 15 years—from 2.6% in 1989 to 1.8% in 2006.

As described previously, the major bond markets cater mostly to the borrowing requirements of their own governments and corporations. However, there are various specialized bonds through which an energy project in a developing country can mobilize funds from these bond markets. An understanding of these specialized bonds requires comprehension of the basic features and the normal restrictions under which a bond is issued and traded.

A bond is issued by a borrowing entity and promises to pay the bondholder a specified annual interest and a specified amount at the time of maturity. A project entity can raise finance through bonds in two ways: public issue and private placement. Procedures for public issue of bonds are broadly the same as for public issue of equity. Public offerings should be approved by the company's board, and the company should file a registration statement with the appropriate regulatory agency (e.g., the Securities and Exchange Commission [SEC] in the United States). This statement contains a lot of detailed information. Afterward, the security can be sold. Unlike an equity issue, the registration of a bond issue includes an indenture between the borrower and a trust company that will act on behalf of all borrowers to insure the terms of the bond. A typical bond indenture includes the following provisions: basic terms of the bond, security or guarantees, covenants, a sinking-fund arrangement, and a call provision. Normally, publicly issued bonds are in small increments (e.g., $1,000 per bond in the United States). After the issue, the bondholder can sell it at any time (prior to maturity) on an organized security exchange. Four variables characterize a bond: its *par value,* also known as *nominal value* or *face value*; *coupon rate,* which is a percentage of par value paid as interest; *maturity date*; and *market value.*

Private placement of bonds is similar to obtaining a loan. The bond is placed with a select number of investors, normally banks, financial institutions, pension funds, and other institutional investors. Raising funds through this mechanism does not require registration with authorities such as the SEC. The advantage of bond placement over a public issue of bond is the saving of time and effort that the company would otherwise spend on registration of the issue. The advantage of private bond placement over a regular loan is that the borrowing company can often get a longer maturity through the bond. The disadvantage of this type of bond as compared to a publicly issued bond or a regular loan is that the interest rate is likely to be higher on a privately placed bond.

A bond, publicly issued or privately placed, may carry an option of converting to the equity shares of the borrowing company at a predetermined price and during a specified time frame. These are called convertible bonds and in effect contain a bond and a *call option.*

A *Eurobond* is a bond issued and traded outside the country in whose currency it is denominated and outside the regulations of a single country. It is often a bond issued by a non-European company in a non-European currency but issued in Europe and sold worldwide. A *Eurodollar bond* is a Eurobond denominated in U.S. dollars. A *Euroyen bond* is a Eurobond denominated in Japanese yen.

A Eurobond is issued by an international syndicate and categorized according to the currency in which it is denominated. For example, consider a Eurodollar bond that is denominated in U.S. dollars and is issued in Japan by an Australian company; the Australian company could issue the Eurodollar bond in any country other than the United States. Eurobonds are attractive financing tools, as they give issuers the flexibility to choose the country in which to offer their bond according to that country's regulatory constraints. They also give the option of the currency in which the company wants to raise money.

Summary 7–4. International Bond Markets

- The major bond markets are the United States, Japan, and the euro area. These markets are primarily used by domestic borrowers, including the host country's government, private corporations, and banks.
- Energy projects in developing countries raise funds through the issuing of bonds by the parent company or by the project company.
- Two instruments, private placement and Eurobonds, are used more often by energy projects in developing countries because of less stringent regulatory and disclosure requirements.

Developing countries issued a total of $130 billion in bonds on the international markets in 2005, double the volume in 2000. The biggest share and growth was related to the issuances from Russia, Poland, Hungary, Brazil, and Mexico. However, more bond issues were on behalf of the governments of these countries than private corporations. The developing-country firms have been less willing to list on the international markets, particularly the United States, because of the availability of more convenient options of raising funds on the domestic debt and equity markets. The disadvantage of raising funds on the U.S. market is now even more evident because of the stringent reporting standards required by the Sarbanes-Oxley Act of 2003.

International Equity Markets

International equity markets can be a significant avenue for the financing of energy projects. There are three major benefits of listing a company internationally: first, it provides access to a larger shareholder base, which reduces the firm's cost of capital; second, it provides access to greater liquidity than is available on the domestic market; third, it increases prestige and visibility and provides a mechanism of signaling the firm's quality. This makes it easier to fund investment opportunities in the future.

Energy companies in developing countries raise funds on international equity markets by using the depositary receipt program. A depositary receipt is a negotiable financial instrument issued by a bank to represent a foreign company's publicly traded securities (which are normally traded on the stock exchange of the country in which the company is located). As such, the depositary receipt enables a company to sell some of its shares in another country. When the depositary bank is in the United States, the instrument is called *American depositary receipts* (ADR); when the bank is in Europe, it is called *European depositary receipts*; and when the bank is in other countries, it is called *global depositary receipts* (GDR).

A project can access the U.S. market and other markets outside the United States either through a private placement of depositary receipts or through a public issue. In the latter case, the issue would have to be registered with the regulatory agency (e.g., the SEC in the United States). However, through the private placement of depositary receipts, the project can raise capital from large institutional investors in the United

States or other countries without registration with the regulatory agency. In recent years, a number of the developing countries, notably India and Brazil, have raised capital in this manner for infrastructure projects.

Depositary receipts have been particularly effective in tapping the U.S. market. There is indeed a demand in the United States for depositary receipts that enable (public and private) investors to buy shares of promising foreign companies. In practice, many U.S. banks will purchase a bulk lot of shares from the foreign company, bundle the shares into groups (to denominate them in U.S. dollars), and reissue them on one of the U.S. stock exchanges. In return, the foreign company must provide detailed financial information to the sponsor bank.

The manner in which the sponsor bank offers the ADR shares to the public varies depending on the level of detailed information that they would provide to the SEC. A level 1 offering takes place when a foreign company does not qualify or wish to be listed on a U.S. stock exchange. This offering will be made available on the over-the-counter (OTC) market and has relatively loose SEC requirements. OTC trading takes place through a network of dealers and brokers, as opposed to trading on a centralized and formal stock exchange, such as the New York Stock Exchange (NYSE) or the American Stock Exchange (AMEX). In contrast, a level 3 offering refers to the formal public offering of ADR shares on a U.S. exchange. This type of offering, while under the SEC's more stringent requirements, gives foreign companies more visibility, more prestige, and better ability to raise capital.

A recent example of raising money through the sale of equity is KazMunaiGas Exploration and Production, a subsidiary of the state oil and gas company of Kazakhstan. The

Exploration and Production Company raised $2.03 billion in 2006 through an IPO. The shares were offered locally, on the Kazakhstan Stock Exchange, and in the form of GDR, on the London Stock Exchange. The issue was split almost equally between the two markets and was fully subscribed.

Summary 7–5. International Equity Markets

- There are three main benefits of listing the shares of a company internationally: access to a greater shareholder base, greater liquidity, and increased prestige of the firm.
- Energy companies in developing countries raise funds on international equity markets by using the depositary receipt program, which itself has various levels of formality and registration requirements.

Specialized Energy Funds

Specialized energy funds are aimed at the mobilization of private sector resources. Establishment of these funds took a lot of effort in the 1990s but has now become a normal business, with rapid growth in terms of number and volume of activities. Many of these funds were initially supported by multilateral and bilateral institutions, as well as the corresponding governments. These support instruments are still available, but the emphasis has shifted to projects that are commercially viable and can be sustained on their own.

Examples of early specialized energy funds are the energy funds of Pakistan and Jamaica. The former fund was created to invest in a private energy project in Pakistan; it was established on the basis of the financial support from the government of Pakistan, the World Bank, USAID, JEXIM, and aid agencies of Italy and France. The latter fund was created to invest in the power sector of Jamaica; it was established with the help of the government of Jamaica, the World Bank, and the IDB.

Today, private funds are formed on a commercial basis with the objective of making aggressive returns through either equity participation or lending to private companies in developing countries. Capital is contributed by institutional investors and large companies. For institutional investors, these funds serve as risk mitigators and facilitators of participation in projects in developing countries. For example, a pension fund in Seattle may not wish to invest directly in a power plant in Thailand because this may seem too risky. However, the pension fund may feel comfortable investing in a global power fund that in turn invests in a diversified portfolio of power projects in several countries, including Thailand.

Participation in private energy funds may be viewed as an opportunity for market development for suppliers of goods and services, such as power equipment manufacturers and engineering firms. At present, there are numerous funds that were introduced by commercial entities to fund energy projects. Many of these funds were created in the Middle East to take advantage of the boom triggered by high oil prices. Examples include the GCC Energy Fund, the Middle East and North Africa (MENA) Infrastructure Fund, and the Infrastructure Growth Capital Fund. These funds are established by well-known international commercial banks, such as HSBC,

Swiscorp, and Deutsche Bank, in conjunction with regional financial entities, such as Dubai International Capital and the Saudi Arabian General Investment Authority.

Still, private funds attempt to maximize their access to resources and minimize their risks by involving governments and multilateral and bilateral institutions. Among multilaterals, the IFC has provided extensive support toward the establishment of infrastructure and energy funds. The African Infrastructure Fund (AIF) has been in operation for about 15 years and has supported a number of energy projects in the Africa region. Similarly, the Asian Mezzanine Infrastructure Fund was established in 1998 to make investment in private projects including power and gas companies. In addition to regional funds, the IFC has supported establishment of country-specific infrastructure funds, such as the IndAsia Fund, which is itself a vehicle for the mobilization of private international capital and the structuring of investment funds for projects in India.

Among bilateral donor agencies, some (e.g., USAID) have always provided support to private funds; others have become very active only in recent years. For example, DFID, in cooperation with other donors, has established a number of vehicles to support private sector infrastructure projects in developing countries. Some of these vehicles (e.g., the Public-Private Infrastructure Advisory Facility [PPIAF] and the Project Development Facility) provide support in preparing private or public-private sector projects. Others (e.g., the Emerging Africa Infrastructure Fund and the Asia Infrastructure Facility) provide financial support in the form of loans or guarantees.

Norfund is another good example. The fund was established in 1997 to promote sustainable and viable ventures in developing countries by providing equity and loans to such ventures. Norfund makes direct investments in private companies and in other funds that have similar objectives. It has a particular interest in the power sector.

Financial Contributions from Project Beneficiaries

Some project beneficiaries are willing and able to make equity or debt-financing contributions because they are interested in providing inputs to the plant or in purchasing its output. Equipment suppliers and contractors represent the most important source in this category. Many manufacturers of large capital equipment have set up captive financing companies for their products. Usually, these companies are specifically designed to generate incremental sales by dealing with less-than-investment-grade credits. Competition often forces such companies to offer very competitive rates not otherwise available on the market. Suppliers of equipment and services can also play an important role in mobilizing funds from the bilateral agencies of their own countries (see chap. 6).

Rating Agencies

The role of the rating agencies

Rating agencies play a major role in access and cost of commercial funding. Lenders, particularly commercial ones, pay considerable attention to the assessments made by rating agencies. These assessments generally indicate the probability of default by the corresponding borrower. In certain financial markets, as in the United States, the public issue of a bond is not permitted if the bond has not been rated.

There are more than 50 rating agencies in the world, the most famous being Standard & Poor's (S&P), Moody's, and Fitch Ratings. Although the major agencies are based in the United States, these agencies are recognized worldwide and have affiliates in many countries.

The rating methodology differs when the agency is evaluating the probability of default for a sovereign, a corporation, or a project company. However, the agencies have adopted systematic criteria to assess the probability of default.

Sovereign ratings measure the ability and willingness of a country to service its debts. S&P's sovereign analysis considers two basic elements—political risks and economic risks. In assessing political risks, S&P considers the stability of the political system, the social environment, and international relations. Orderly succession in political leadership, the system's adaptability to changing circumstances, and the extent of social consensus on basic questions contribute to a strong sovereign rating. The degree of integration into multilateral trade and international financial systems is further regarded as an incentive for a government to honor its foreign obligations.

S&P's economic analysis reviews a government's external financial position, balance-of-payments flexibility, and management of the economy, as well as the country's overall economic prospects. The primary aim of this analysis is to assess the debt burden carried by the country and the likely evolution of the country's debt-servicing capacity.

At the project level, ratings assess the following:

- Likelihood of default by the project company on its timely payment of (principal and interest) loans
- Nature of the project and provisions that are put in place to avoid such defaults
- Protection of the lenders' interest in the event of default, bankruptcy, or other arrangements that may affect the creditors' rights

Project-level ratings consider six major risks:

- Contractual
- Construction and operations
- Competitive market exposure
- Legal structure
- Counterparty
- Financial strength

Analysis of the contractual risks includes an evaluation of the contracts that have been implemented and consideration of whether they adequately mitigate the operational risks of the project. Examples of such contracts are power purchase agreements and concession agreements. Assessment of contractual risks also includes an evaluation of the collateral agreements.

Construction and operations risks encompass both preconstruction and postconstruction risks. Preconstruction risks are addressed through the evaluation of engineering and design of the project. Assessment of postconstruction risks is focused on the ability of the project to operate and maintain the facility.

Competitive market exposure entails the evaluation of the potential for new entrants into the industry. Such an evaluation also establishes the project's competitive advantage in its corresponding sector.

Analysis of the legal structure encompasses a review of the entities connected to the project. For instance, sponsors and suppliers are evaluated.

Counterparty-risk analysis includes an evaluation of raw material suppliers, off-take purchasers, and EPC contractors.

Finally, and most important, the financial strength of the project company is evaluated, to assess its ability to deal with changes in the financial environment. An important factor is the debt-service coverage ratio (DSCR), which indicates whether a project company is likely to be able to make timely interest payments.

Traditionally, rating agencies assumed that before a government defaulted on its foreign obligations, it would freeze foreign payments by corporate borrowers. Hence, the rating agencies would not usually assign project ratings that exceeded sovereign foreign currency ratings. However, experience in the 1990s indicated that governments do not necessarily stop foreign payments by corporations when they default on their own bonds.

Agencies have since modified their approach to better reflect the changing environment. Even using the stricter criteria now in place, rating agencies have occasionally rated a project above the sovereign rating when firms are considered to be a lower credit risk than the government. In these cases, even if the government defaults, the expectation is that it will not automatically force the project to default.

Summary 7–6. The Role of Rating Agencies

- Rating agencies provide a recognized benchmark that investors and lenders consider when deciding whether to fund a project.
- There is a large number of rating agencies. The ratings most frequently used are S&P's and Moody's.
- The rating methodology contains different factors for assessment of credit rating for sovereign, corporate, and project issues. Rating agencies adapt to changing market conditions and environment but attempt to keep the ratings stable and consistent.

Interpretation of ratings

S&P's rating spectrum runs from AAA to DDD and is interpreted as follows:

- *AAA bonds* are judged to be of the best quality. These *gilt-edge bonds* carry the lowest degree of investment risk. Interest payments are protected by a large or an exceptionally stable margin, and principal is secure. Although the various protective elements are likely to change, changes are unlikely to impair the fundamentally strong position of such issues.
- *AA bonds* are judged to be of high quality by all standards. Together with the AAA group, they comprise what are generally known as *high-grade bonds*. They are rated lower than the best bonds because margins of protection may not

be as wide as in AAA securities, fluctuation of protective elements may be of greater amplitude, or other elements may be present that make the long-term risks appear somewhat greater than in AAA securities.

- *A bonds* possess many favorable investment attributes and are considered to be *upper-medium-grade obligations*. Factors giving security to principal and interest are considered adequate, but elements may be present that suggest susceptibility to impairment in the future.
- *BAA bonds* are considered *medium-grade obligations*; that is, they are neither highly protected nor poorly secured. Interest payments and principal security appear adequate for the present, but certain protective elements may be lacking or may be characteristically unreliable over any great length of time. Such bonds lack outstanding investment characteristics and in fact have speculative characteristics as well.
- *BA bonds* are judged to have speculative elements; that is, their future cannot be considered to be well assured. Often, the protection of principal and interest payments may be very moderate and, therefore, not well safeguarded during both good and bad times over the future. Uncertainty of position characterizes bonds in this class.
- *B bonds* generally lack characteristics of a desirable investment. Assurance of either principal and interest payments or maintenance of other terms of the contract over any long period may be small.
- *CAA bonds* are of poor standing. Such issues may be in default, or elements of danger may be present with respect to principal or interest.

- *CA bonds* represent obligations that are speculative to a high degree. Such issues are often in default or have other marked shortcomings.
- *C bonds* are the lowest-rated class of bonds, and issues so rated can be regarded as having extremely poor prospects of ever attaining any real investment standing.
- *D bonds* are an issue that is already in payment default or whose issuer has filed for bankruptcy.

Moody's rating system has similar interpretations. The correspondence between Moody's and S&P's ratings is as follows:

S&P	Moody's
AAA	Aaa
AA	Aa
A	A
BBB	Baa
BB	Ba
B	B
CCC	Caa
CC	Ca
C	C
D	D

PART III

DESIGNING AN ACCEPTABLE PROJECT PACKAGE

The ultimate objective of the project package is to convince investors and financiers that funds put into the proposed project will result in a sufficiently high and secure return. Thus, two intertwined questions—namely, the project return (or, for senior lenders, cash flow) and its corresponding risks—represent the cornerstone of project preparation, around which knowledge about several other areas is built. These other areas include technical soundness of the project, environmental concerns, market potential, and political environment and are investigated to ensure that the project return remains adequate, stable, and safe (most major lenders have declared that they consider sustainable development and environmental management independently and in addition to adequate and safe return).

Although many energy projects are built to serve the domestic market, they are always linked to international markets. The general framework for analyzing the viability of a project involves the following four steps:

- Potential demand for project output should be assessed by using consumption forecasts by various customer groups and analyzing existing and forthcoming supply capacity. In the event that part or all of the project output is aimed at exports, the analysis should cover the specific markets targeted by the project.
- In relation to the previous step, the evolution of energy prices, the tariff structure, and all relevant laws and regulations should be analyzed to arrive at a realistic projection of future prices for output of the energy project.
- International norms and market conditions should be reviewed to establish appropriate references for capital and operations and maintenance (O&M) costs. These could include average prices of capital equipment, fuel prices, and interest rates.
- Local conditions affecting capital and O&M costs should be analyzed, and adjustments should be made according to international average costs. These could include a country multiplier, physical conditions (e.g., plant site, conditions

and remoteness, and pipeline terrain), environmental constraints, laws and regulations, and local input conditions (e.g., access and communications infrastructure, as well as fuel and labor markets).

There are many areas of overlap among these four steps. However, the overall aim of the analysis is to use results of the first two steps to arrive at the likely stream of project revenues, as well as possible variations in these revenues. In the same way, the third and fourth steps establish a stream of expected project costs, as well as possible variations in these costs.

The revenue and cost streams, along with the range of their variations, form the basis of the risk-reward profile of investment in the proposed project. If the resultant risk-reward profile is attractive, then the ownership and financing structures are designed on the basis of the analysis of availability of funds from public/private, domestic/foreign, and international sources.

A systematic approach to project preparation involves

- Analyzing the international and local business environment
- Establishing project viability
- Structuring the financing package

Accordingly, part III of this book contains a chapter on each topic. In addition, because environmental concerns have become very important in the preparation of energy projects, the final chapter is devoted to this matter.

8 Analyzing The Business Environment

Analysis of the business environment provides a reasonable assessment of the chances that a proposed project will achieve its economic and financial objectives. In practice, this assessment boils down to the establishment of a basis for assumptions made regarding streams of project costs and benefits.

Although various elements of the business environment are intertwined, the analysis normally focuses on three general categories:

- Project-specific business environment
- Host country's business environment
- International business environment

The Project

The project-related analysis provides an overview of the institutional, technical, and cost aspects of the project (fig. 8–1). The institutional aspects include

- Company establishment and ownership—for example, background, purpose, and shareholdership
- Organization and management—for example, composition of the board of directors, relationship with the chief executive officer, and company organization

- Employment and staffing—for example, acquisition of expertise, employment of local staff, and promotion of indigenous expertise
- Background information about project sponsors—for example, ownership structure, business objective, management, capital, and financial and operating performance of each sponsor

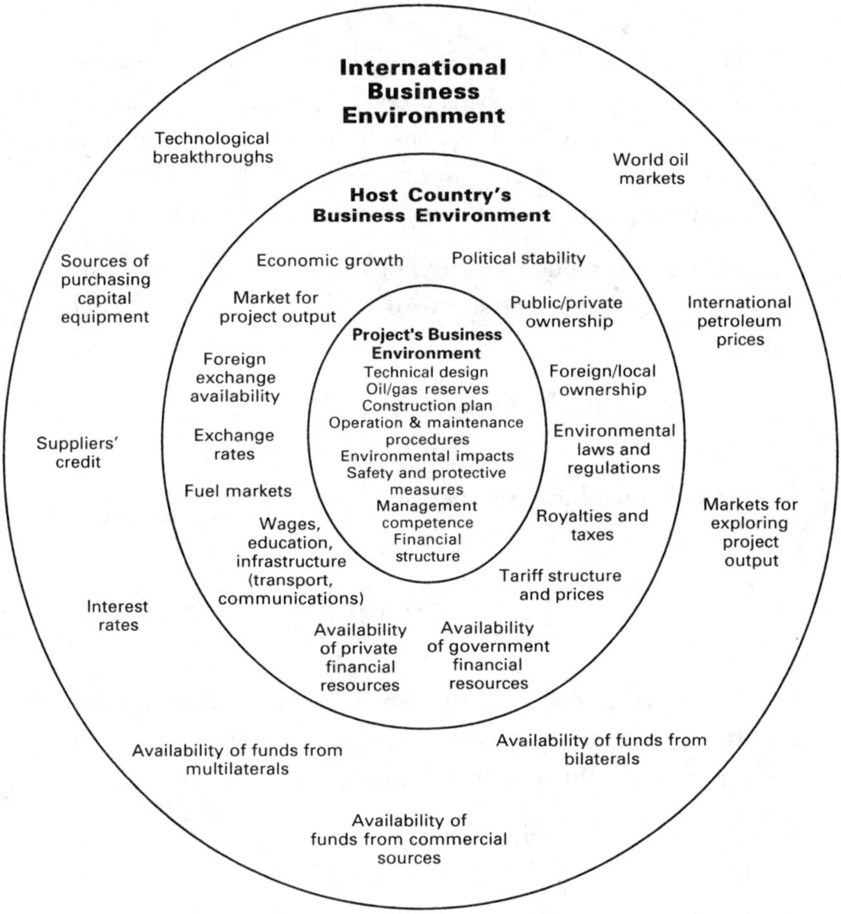

Fig. 8–1. Major interactions among the project environment and the business environment

The technical analysis of the project covers all important issues related to design, construction, and operation of the project (fig. 8–2). The following are the major items:

- The project description, including assurances that the design, engineering, procurement, and construction of project facilities will be in accordance with sound professional engineering practices and will be based on company guidelines in conjunction with relevant applicable international codes and standards and local legal and regulatory limitations. Performance and reliability completion tests should be explained properly.
- The project site, including description of access to transportation facilities, water and wastewater facilities, human habitation, fauna, and flora. This description is important in determining project costs and environmental impacts.
- The plant operating philosophy and procedures, including the degree of safety, reliability, flexibility, and efficiency. Safety is a major concern in all energy projects. Accordingly, this area receives great emphasis at the stages of design and construction. Nevertheless, minimizing risks also requires effective working and operating procedures, use of preventative maintenance practices, strict adherence to instructions and guidelines, and adequate incentives to facilitate these procedures. Reliability, flexibility, and efficiency are largely related to project design but can be enhanced through effective operating procedures.
- Facilities maintenance, including assurance of regular maintenance, as well as planned shutdowns for major scheduled maintenance.
- Safety and protective measures. These are discussed extensively in the chapter on project environmental assessment (see chap. 11).

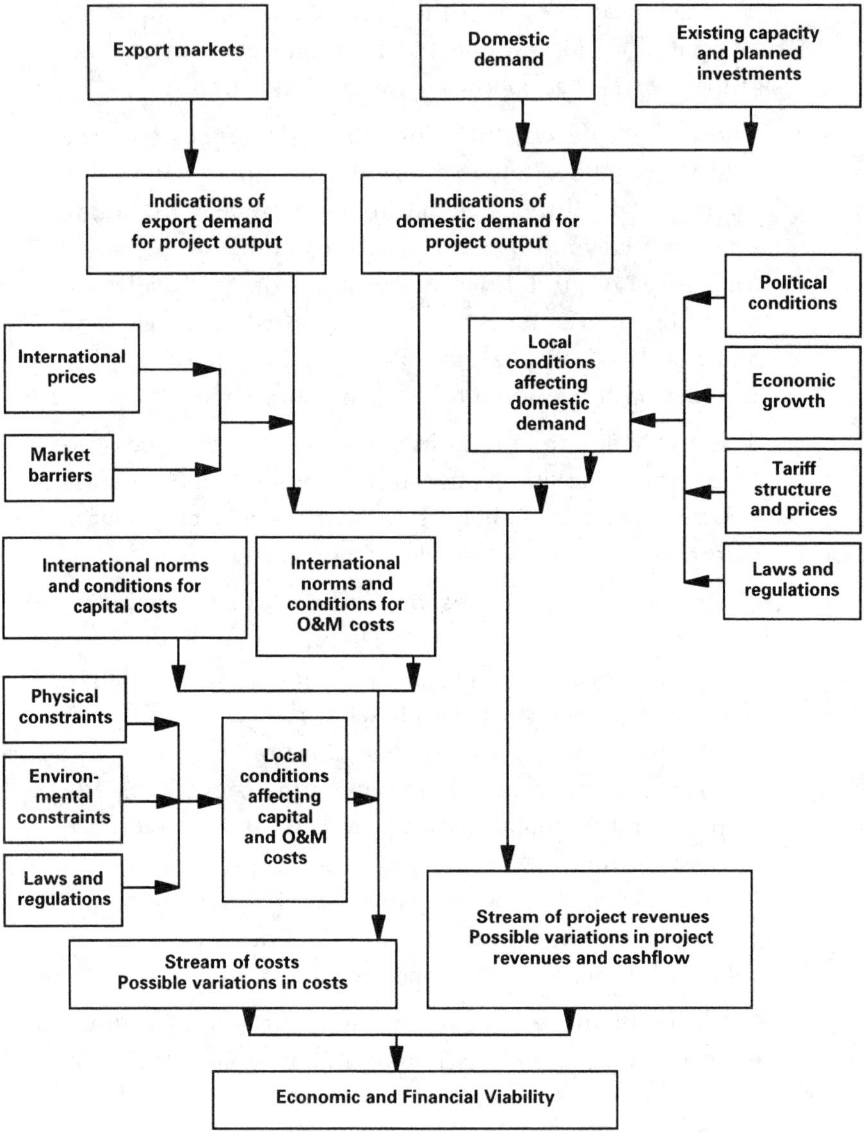

Fig. 8–2. Factors affecting project viability

- Auxiliary services, including electric power, air and water supply, cooling-water systems, and fuel systems needed to support the main plant or facility. The design features and operations standards of each facility should be explained sufficiently.
- Agreements on and arrangements of technical assistance, to ensure that required expertise is available during project construction and operation. In power projects, technical assistance is provided through international consultants or advisors. In oil projects, foreign partners provide substantial assistance. The project document should include terms of reference for all technical assistance services, particularly if a new company is to undertake the project.

The final discussion in the project section should cover expected project costs, procurement, and implementation arrangements. Project costs are normally estimated on the basis of previous experiences in the country, similar projects in other countries, and conditions related to the location of the proposed project. In addition, when the project may be implemented through a turnkey contract, formal and informal quotations may refine cost estimates.

Although project costs depend on time, location, and business environment, average cost numbers have emerged from international experience (table 8–1). These rule-of-thumb figures, known as *U.S. Gulf estimates,* are often used by experts to arrive at the first-cut estimates for each type of energy project. The U.S. Gulf estimates must then be adjusted by a country multiplier. The multiplier varies between 1.1 and 1.25, depending on the country and the region.

TABLE 8–1. RULE-OF-THUMB AVERAGE COST FIGURES

- Upstream development costs vary substantially depending on field characteristics and the water depth, if offshore. Consequently, no generally accepted average cost figures are available.
- Onshore pipeline construction costs vary between $20 and $30 per inch-diameter per meter, depending on terrain and other characteristics. A figure of $25 per inch per meter is normally used for first-cut estimates of pipeline investments inclusive of material and construction costs. Offshore pipelines may cost up to 100 per cent more than onshore pipelines.
- The average cost of transporting gas depends on the volume as follows:

Volume (MMCFD)	Average Cost ($ per MMBTU per 1,000 km)
1,000	1.00
2,000	0.80
3,000 and above	0.60

  Offshore transportation may cost up to twice the above numbers.
- LNG liquefaction plants cost about $300 per ton of plant capacity. The average cost of liquefaction is about $1.30/MMBTU and expected to decline to $1.00/MMBTU in the new plants. The average cost of regasification is about $0.30 /MMBTU.
- Refinery construction costs depend on the degree of technology sophistication and operational flexibility. For simple hydroskimming refineries, average costs are roughly estimated as follows:

Capacity (b/d)	25,000	60,000	120,000	200,000
Capital cost ($/bl of capacity)				
Refinery without infrastructure improvement	3,300	2,600	2,100	2,000
Refinery with infrastructure improvement	4,200	3,400	2,600	2,300
Greenfield refinery	5,800	4,150	3,650	3,425

- Construction costs of power plants is estimated at

  $1,250/kW for coal-based power plants, including flue-gas desulfurization (FGD) facilities.

  $1,500/kW for integrated coal gasification combined-cycle plants.

  $800/kW for oil-fired steam plants.

  $600/kW for gas-based combined-cycle plants.

  $2,000-3,000/kW for nuclear plants.
- Investment costs of power transmission and distribution vary substantially depending on population density and stage of the development of the existing system. Some rough approximations can be made based on $130 to $160/kW for transmission and $190 to $225/kW for distribution facilities.
- All of the cost figures above are in 2005 dollars. A country multiplier between 1.1 and 1.25 is applied depending on the country and the region.

For the linkage of costs to financing needs, it is necessary to separate costs into foreign and local currency. In general, local partners, including state entities, feel most comfortable with providing the local currency component of the project cost while leaving the rest for foreign equity and debt financing.

It is also necessary to separate costs by items that can be procured independently. The actual procurement arrangement cannot be finalized until the financing package is structured. Indeed, procurement for each project component is often chosen or modified in accordance with the requirements of the agency or company providing finance for acquisition of that component. A normal practice is to acquire the main equipment from countries that have substantial suppliers or that provide bilateral credit. The procurement arrangement is then made according to the requirements of the provider of the credit.

The project construction schedule should include completion dates for the main project components, as well as for the auxiliary systems, input deliveries, and off-take and marketing of the output. To show that the schedule is realistic, project sponsors cite their experience in the country. If experience in the host country is lacking, they can cite their experience in other countries, as well as other people's experience in the host country. Again, if a turnkey contract is planned, completion dates should have been discussed tentatively with prospective contractors.

Host Country's Business Environment

Analysis of the host country's business environment should highlight useful indications of the project's costs and benefits, based on the current status and foreseeable developments. The following are the major areas to be addressed:

- The political system and circumstances, to assess political stability, security of investment, attitude toward private and (particularly) foreign investors, and adherence to laws, regulations, rules, and agreements
- Economic conditions, to assess likely changes and possible impacts on market demand and price of project output, as well as on the prices of project inputs (including labor costs) and the foreign exchange rate
- Energy sector development patterns, to forecast the overall demand/supply balance and associated physical, financial, and institutional constraints on the supply of various forms of energy
- Subsector issues, to assess the eventual market demand for project output and issues that project sponsors have to face during the construction and operation phases

THE POLITICAL SYSTEM

The political analysis assesses political stability and the degree of government intervention in the economy. In terms of the overall political matters, it is important to provide concise information about

- The relative powers of the legislature versus the executive branch
- The center(s) of power in the executive branch
- Turnover in the executive branch

- The significance of opposition parties and grassroots participation
- The government's manner of handling the opposition
- The country's attitude toward and participation in regional and international cooperation

With regard to government intervention, the analysis should explain the relevant laws, regulations, and track records in

- Issuing licenses for major investments to private and foreign investors
- Intervening and exercising control when private and foreign investors are involved
- Restricting payments of dividends to private shareholders
- Procuring goods and services when the government is involved in financing, guaranteeing, or otherwise sponsoring foreign borrowing

Economic conditions

Macroeconomic conditions can be analyzed in terms of the strength of domestic markets, the levels of costs and prices, the country's foreign-debt position, and the availability of funds on domestic capital markets. To assess the economic factors affecting market demand, an analysis of historical and projected trends should be undertaken for GNP, national consumption, and GDP growth rate and sectoral composition (fig. 8–3).

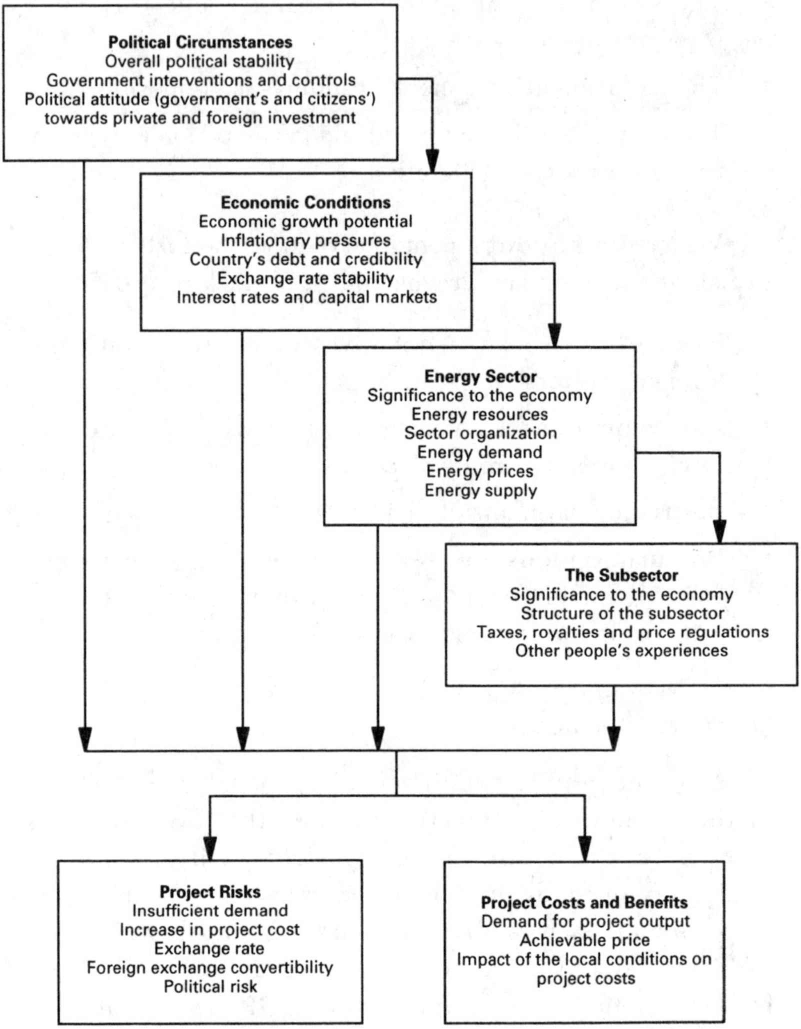

Fig. 8–3. Analysis of host country's business environment

The economic indicators related to costs and wages include inflation indexes (e.g., GDP deflator and consumer-price index), wage indexes, deflators for fixed capital formation and investment expenditures, and formal and informal interest rates. It is also necessary to assess the government policy and track record in combating inflationary pressures.

The foreign-debt position of a country is an important determinant of the country's official and unofficial credit rating. The amount and evolution of foreign debt should be explained by examining import and export trends, capital flight, and the resultant accumulated foreign debt. It is also important to determine the portion of the debt that is in the form of public and publicly guaranteed debt and the portion that is in the form of private nonguaranteed debt.

Finally, discussion of economic conditions should briefly address the financial sector. This includes the banking system, the domestic bond market, other institutional investors, and any precedents for mobilizing funds for major projects through public offerings or private placements of equity shares or bonds.

The Energy Sector

By analysis of the energy sector, a framework can be developed for assessment of competition among various forms of energy and of the advantages, constraints, and resource requirements for various energy supply options. Energy sector analysis should cover the following areas:

- Significance of the energy sector to the economy. This significance is assessed both in terms of the ratio of the energy sector's output to total GDP and in terms of the percentages of national investment and public budget used in the sector.

- Energy resources, including recoverable amounts of oil, gas, and coal, as well as hydroelectric power potential.
- Sector organization and management. The institutional structure should be discussed, to determine the relative shares of public and private sector participation in the supply of energy and to understand how the government interacts with the energy sector. The discussion should include the roles that various government agencies play in the energy sector.
- Energy demand, including an analysis of historical patterns and forecasts for 15–20 years. The analysis should be based on the sectoral (residential, commercial, and industrial) composition of energy demand and should be broken down by type of primary energy (oil, gas, hydroelectric, coal, and nuclear) and final energy (petroleum, gas, and electricity).
- Management of energy demand, including regulation of energy prices and promotion of energy conservation.
- Current capacity and sources of supply, including installed capacity for production of oil and gas and generation of power, as well as sources of imports of crude oil or petroleum products.
- Constraints and issues in expanding energy supply, including possible insufficiency of domestic energy resources, heavy burden of energy investments on public budget, subsidies and cross-subsidies in energy prices, and corresponding distortions and environmental constraints.

The relevant subsectors

The subsector analysis should include

- The significance of the subsector to the economy. Each subsector—petroleum, gas, and power—plays a special role in the economy. The petroleum sector is important because it yields royalty and tax revenues. The power sector is important because it facilitates economic growth and development. The gas sector has some of both features and may provide environmental benefits.
- The structure of the subsector. For each subsector, it is important to assess the extent and potential for competition. The current sector structure—for example, the extent of integration, the number of companies, and the share of the largest company or companies—is an important indication of potential competition.
- The government role. It is said that governments do three things: tax, spend, and regulate. Within each of the energy subsectors, these functions are normally determined in the legislative and regulatory framework. The legislative framework determines, through specific laws (e.g., petroleum and gas acts), provisions regarding ownership and operation of energy sector resources and facilities. The regulatory framework includes procedures for implementing legislative provisions. If the legislative and regulatory frameworks are well established, then the important parameters of a proposed project can be determined. However, most developing countries have no comprehensive legal and regulatory framework in place. Therefore, many key relationships have to be worked out through specific agreements. Nevertheless, an assessment should be made of applicable taxes, achievable prices, and their foreseeable evolution.

- Relevant experience of others. Experiences in the subsector (e.g., projects undertaken by oil companies or private power producers) are analyzed to identify behaviors characteristic of past success and to outline lessons that could facilitate preparation and implementation of the proposed project.
- Analysis of market demand/supply capacity. This is the core of the subsector analysis and should cover the following four topics:
  - Historical perspective of demand/supply
  - Forecast of demand over 15–20 years, under various assumptions about economic growth and energy prices
  - Planned investments
  - Projected supply deficit

International Business Environment

The international business environment can have varying degrees of relevance depending in particular on the sources of project inputs and the markets for project output (fig. 8–4). The main areas of analysis include

- World oil markets, international energy prices, and, if applicable, markets for export of project output
- Sources of capital equipment, interest rates, and availability of funds from multilateral, bilateral, and commercial sources, including suppliers' credit

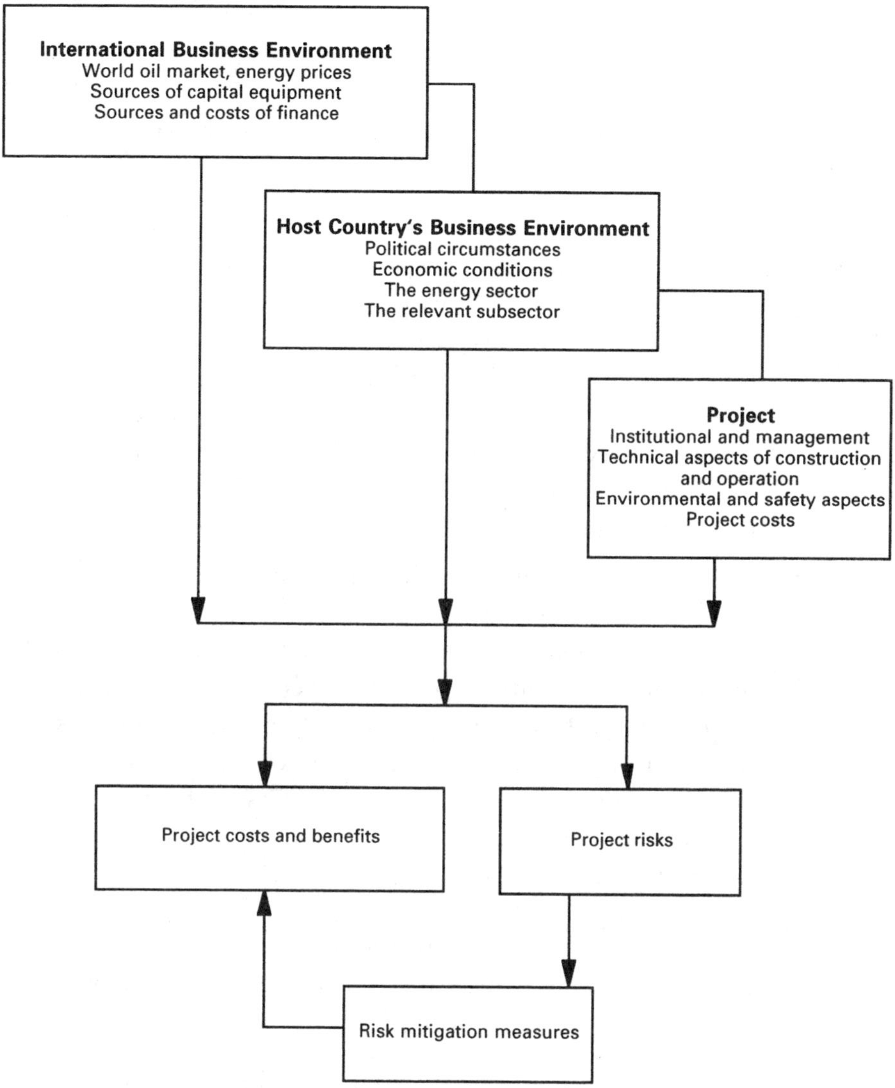

Fig. 8–4. The impact of the international business environment on project viability

WORLD OIL MARKETS AND INTERNATIONAL ENERGY PRICES

The depth of analysis of international energy markets depends on whether the project output is fully or partially aimed at export markets. If export is not relevant, then objectives of the analysis are as follows:

- To provide an outlook of international energy markets and the evolution of energy prices. These prices are normally necessary for project analysis because even if project output is not exported, the domestic price would have some correlation with the international price. In addition, in certain projects (e.g., power generation), the fuel may have to be procured on international markets or bought domestically at prices related to international prices.
- To assess the correlation between domestic and international energy prices. Domestic energy prices have been subsidized in many developing countries. The extent of subsidy varies greatly among countries, but it has been generally reduced during the past two decades. Moreover, most countries plan to remove overall price subsidies, although some cross-subsidies will remain. A further issue affecting the relationship between domestic and international energy prices is the status of price deregulation.

In the event that output of the proposed project is destined for export, the analysis of international energy markets will have to be much deeper, covering

- Specific geographical markets targeted for sale of project output.

- Expansion plans and comparative advantages of suppliers active in targeted markets.
- Legal and regulatory issues relevant to imports.
- Justification that project output can be successfully marketed at the expected price level. Often, a long-term take-or-pay contract with credible purchasers needs to be secured. The arrangement could also be acceptable if the take-or-pay contract is made with a reputable trading house.

Sources of capital equipment

The energy equipment supply market is quite competitive. Furthermore, competition is not limited to price and quality of service. Equipment suppliers often take a flexible approach and do whatever it takes to win business. They could become partners in the proposed energy project, finance purchases, or influence others who finance the project.

Analysis of the equipment supply market covers

- Identification of suppliers with compatible technology and track records in providing satisfactory service during plant construction and operation.
- Examination of the corresponding countries' bilateral facilities, including aid, soft loans, and guarantee programs. It is also important to review the decision-making processes involved in these programs. Of particular interest would be the influence that each supplier would have in such decisions.
- The financial capacity of equipment suppliers and their normal practice in providing financial assistance in the form of partnership, loan, or credit.

SOURCES AND COSTS OF FINANCE

A logical extension of the previous analysis is the review of capital markets. This analysis should cover

- Availability of funds suitable to the proposed project. This would draw on experience of financing similar projects and assess the preparedness of multilateral, bilateral, and commercial financiers to support the proposed project.
- Cost of capital—for example, evolution of interest rates for each source or method of financing.

THE USE OF RISK MITIGATION INSTRUMENTS

Analysis of the business environment reveals a host of risks that could endanger a proposed project (table 8–2). Before analyzing project costs and benefits, sponsors should utilize available guarantee or insurance schemes to manage these risks. Many methods and a variety of sources are available to guarantee virtually all significant transactions.

TABLE 8–2. TYPES OF PROJECT RISKS AND PARTIES INFLUENCING RISK

Type of Risk	Parties Influencing Risk
Commercial Risk	
CONSTRUCTION RISK	
Cost overrun	Contractor
Completion delay	Contractor
Increased financial costs	Banks, other lenders
OPERATION RISK	
Unsatisfactory plant performance	Contractor, operator
Excessive maintenance costs	Operator
Fuel supply or fuel cost for power plants	Fuel supplier
Insufficient reserves for oil and gas projects	Sponsors
REVENUE RISK	
Insufficient sale volume	Purchasers of project output
Low prices	
Political Risk	
REGULATORY RISK	
Changes in regulatory regime including price, environmental obligations, and so on	Government
TRANSFER -OF-PROFITS RISK	
Foreign exchange inconvertibility, restrictions in transferring funds	Government
OTHER	
Expropriation, war, civil unrest	—
Force Majeure Risk	
Natural disasters	—

In the absence of the aforementioned guarantees, project sponsors themselves must become guarantors of all risks (see table 8–2). This creates a number of problems. First, the risks may simply be excessive for sponsors who are primarily in the business of building projects, rather than taking risk; risk taking is now a business of its own, which operates within a very sophisticated array of parameters. Second, even if sponsors are willing to assume all the risks, lenders may not feel sufficiently secure to finance the project.

Project sponsors have three main incentives for obtaining third-party guarantees:

- Shifting some of the project risks to other parties
- Reducing the impact that debts of the project company have on their own balance sheets and credit ratings
- Persuading lenders to extend financing to the project

Some minor guarantees are available from commercial guarantors such as banks, insurance companies, and investment companies. These are normally in the form of letters of credit, which are provided on the basis of the credibility of project sponsors. However, the main potential guarantors are parties with interests in some aspect of the project. These include

- Governments and development agencies interested in the economic impact of the project. They may be persuaded to provide guarantees against political risk, problems of foreign exchange convertibility, and certain aspects of market risk—by, for example, underwriting take-or-pay contracts. In addition, a large number of agencies sponsored by the governments of industrialized countries provide bilateral insurance against political risk of investment and finance committed by their own nationals in developing countries.

- Suppliers of equipment and services—including contractors—who want to sell to the project company during the construction phase. They will normally guarantee the cost, timeliness, and performance of their equipment or service.
- Suppliers of raw materials and fuels who want to sell to the project company during its operational phase. They are normally willing to guarantee the delivery and the price (or a price formula).
- Consumers of the output of the proposed project. They may be willing to sign a take-or-pay contract for purchase of the output. This type of arrangement is viewed as an indirect guarantee and is taken into consideration by lenders and credit-rating services.

Guarantee arrangements relating to commercial risks (i.e., project completion, cost overrun, delay, fuel supply, O&M costs, and market demand) are usually agreed during project preparation, between project sponsors and other participants. Mitigation of risks requires considering all possible sources of guarantee and choosing a mix of instruments to enhance prospects for project financing at the lowest possible cost to project sponsors.

The guarantee instruments (fig. 8–5) for dealing with construction risks fall into the following categories:

- *Sponsors' completion guarantee.* The construction and startup phases of the project constitute an important period of risk in project financing. Financiers prefer to receive various types of assurance that construction risks have been managed. The first and most important assurance comes when project sponsors themselves provide a project completion guarantee, which commits the sponsors to completing the project within a certain time period and to providing funds to pay all cost overruns. However, other guarantees—by contractors, equipment suppliers, and so on (as described later)—are normally sought, to protect the project sponsors and to provide financiers with additional assurances that project completion risks have been managed.
- *Lump-sum turnkey (LSTK) contract.* LSTK engineering, procurement, and construction (EPC) contracts are used to ensure that the project is completed on time, within budget, and with acceptable standards of operation. These contracts have stringent reward and penalty clauses that protect sponsors against cost overruns and operational deficiencies.
- *Bid bond.* At the outset of a project, contractors are invited to bid for the project. A bid bond is required of bidders on a contract to ensure that each bidder is serious (i.e., would accept the contract if awarded).
- *Performance bond.* A performance bond provides additional funds in the event that the contractor fails to perform for any reason. The existence of such a bond

is also an endorsement of the credit and confidence of the guarantor in the ability and professional standing of the contractor.

- *Advance-payment guarantee.* An advance-payment guarantee assists the contractor in purchasing and assembling the materials, equipment, and personnel necessary to begin construction, to meet the requirements for receipt of progress payments under the contract.
- *Retention bond.* It is common practice for the sponsors for whom a project is being built to retain or hold back a portion of the progress payment that would otherwise be due, to provide a fund that covers unforeseen expenses caused by any contractor mistakes in construction. Because most contractors prefer to receive the progress payments as quickly as possible, they substitute a retention bond for the amount of the funds retained, to receive immediate payment.
- *Maintenance bond.* The purpose of the maintenance bond is to provide a source of funds for the correction of construction or performance defects discovered after completion of construction. Typically, the performance bond and the retention bond are converted to maintenance bonds on completion of the contract.
- *Financial derivatives.* Financial derivatives, such as swaps and options, can be used to hedge against unprotected increases in financial costs, particularly those caused by an increase in interest rates or an adverse movement in the exchange rate. Both of these factors can significantly affect the cost of the project, especially if sponsors have borrowed funds at floating rates and in a single currency. Interest rate swaps and caps, as well as currency swaps and options, may be used to deal with these risks.

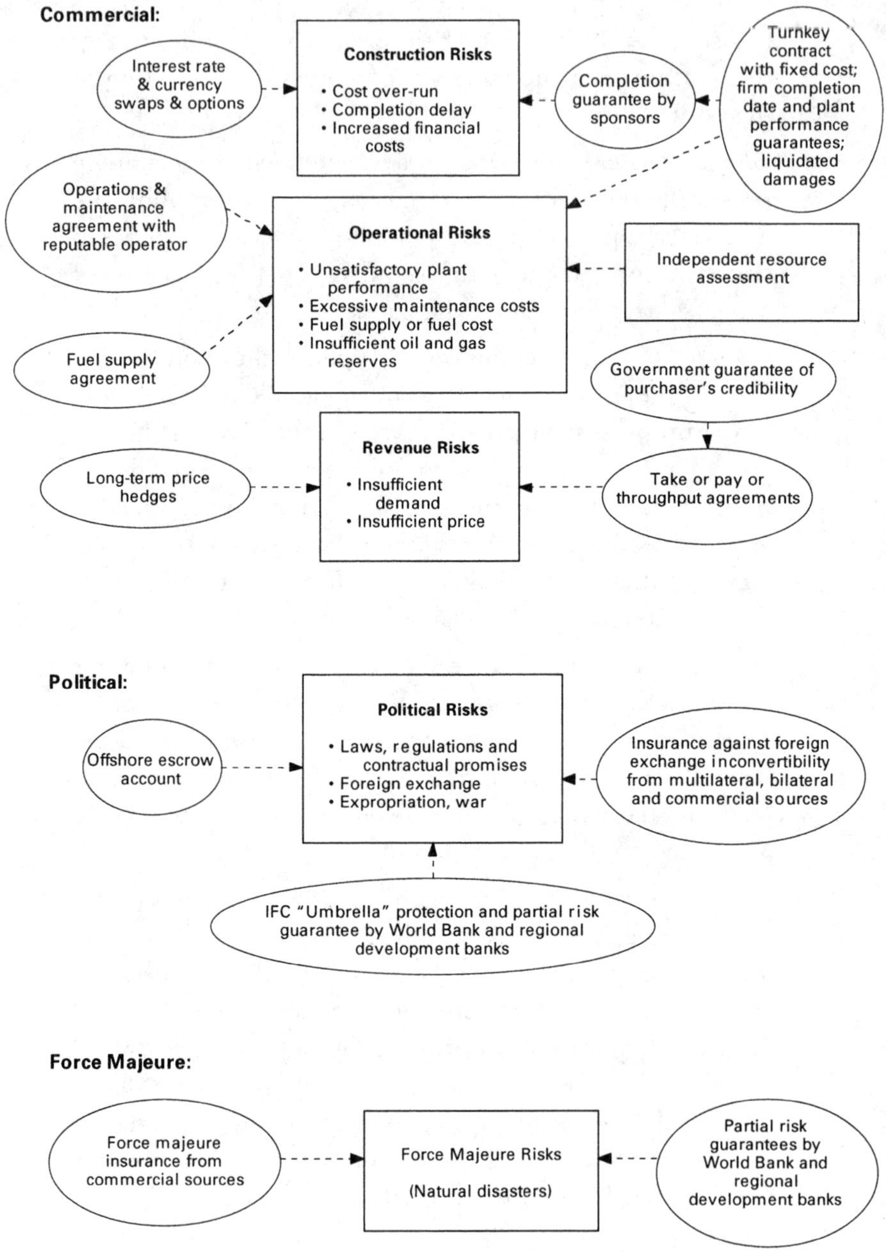

Fig. 8–5. Risk mitigation and guarantee instruments

In addition to well-designed EPC contract protection against operating deficiencies, guarantee instruments for dealing with commercial aspects of operational risks include the following:

- *Put-or-pay contract.* Any unforeseen change in availability and cost of energy, raw materials, or product will affect the performance and profitability of a project. Most projects hedge against these risks by utilizing put-or-pay contracts (also called *supply-or-pay contracts*). Put-or-pay contracts are provided by suppliers of energy, raw materials, or products to projects needing assured supplies of such inputs over long periods and at predictable prices, to meet production cost targets. The put-or-pay obligor must either supply the energy, raw material, or product or pay the obligee (i.e., the project company) the difference in costs incurred in obtaining the input from another source.
- *Take-or-pay contract.* A take-or-pay contract is an unconditional contractual arrangement between the project sponsor and the customer. The contract obliges the customer to make periodic payments in the future in return for fixed amounts or quantities of products, goods, or services at specified prices. The obligation to pay is unconditional and must be fulfilled regardless whether the service or the product is used by the customer. As the cost of service may rise over time because of inflation, payments are usually subject to escalation. The take-or-pay obligor can protect its interest by retaining rights to take over the project in the event of failure by the supplier to perform. Any such takeover would be subject to the take-or-pay obligor's assuming or paying the debt used to finance the project. Examples of take-or-pay contracts can be found in almost all segments of the energy market.

- *Throughput contract.* The equivalent to take-or-pay contracts for projects that provide services such as power transmission, power distribution, oil pipeline transportation, and refining is a throughput contract. Lenders regard throughput contracts as a guaranteed source of income; the guarantee is unconditional and lasts for the life of the loan. The obligor pays regardless whether the service is used. This type of obligation is sometimes called a *tolling agreement,* a *cost-of-service tariff,* or a *deficiency agreement.*
- *Escrow account.* An escrow account is not a guarantee but a means of payment, a way to put aside the money needed to repay a debt. The escrow account is a special account, often outside the host country, to which a certain portion of the project revenues are channeled. The account is managed by an agent in accordance with a specific agreement. If all project revenues are deposited in the account, then the escrow agent disburses the funds with a structured priority, first covering the operating cost and then covering debt service before remitting the remainder to the project company. The escrow account agreement terminates when all debt obligations have been fully paid.

The guarantee instruments for dealing with political (country) risks are more complex because of difficulty in precise definition and assessment of resultant defaults. In a broad sense, political risks include

- Currency risks—for example, inconvertibility, devaluation, and restrictions on currency imports and exports
- Taxes and duties—for example, increased taxes on property, production, income, and profits, as well as increased import or export duties

- Labor risks—for example, changes in laws and regulations dealing with work permits for imported labor, labor unions, and labor compensation
- Government-intervention risks—for example, local government and federal government interference or harassment through licenses, regulations, police, and military
- Losses from expropriation, nationalization, confiscation, war, or revolution

No guarantee instrument covers all of the aforementioned risks. Often, project sponsors combine several guarantees to ensure sufficient protection against political risks. For example,

- Guarantees by the host government are often needed to assure the project company that the government will take measures to protect or enhance the interest of the project, provided that it is within the government's control and that the company functions within the country's legal framework. Host-government guarantees against political risk can take many forms, including concession agreements covering some of the risks outlined previously, coverage against expropriation or nationalization, and guarantee by the central bank of the host country against risks such as currency and foreign exchange restrictions.
- MIGA, an affiliate of the World Bank, provides investment insurance against political risk in developing countries.
- The IFC, another affiliate of the World Bank, provides various types of comfort to other investors and financiers through its umbrella arrangement.

- The World Bank itself may provide partial guarantee against the risk arising from nonperformance of sovereign contractual obligations and from the force majeure aspects of the project.
- Regional development banks (e.g., IDB, EBRD, and ADB) provide various types of guarantees against political risk.
- Most export-import banks of industrialized countries provide various types of guarantees against commercial and political risks.
- Most industrialized countries now have specialized government-sponsored agencies that guarantee investments and financing of their own nationals in developing countries.

Sources of Information for Preparation of Energy Projects

Information for the preparation of an energy project can be obtained from numerous publications and Web sites. This section names a few sources that are good starting points for compiling data, understanding policy issues, and developing future outlooks.

Country information

Often it is easier to compile information about a country's political and economic conditions from sources outside, rather than inside, the country. At the least, one should use such outside sources to do his or her homework before committing to a project in a developing country.

The most convenient starting point for the study of political and macroeconomic trends is to review reports by the Economist Intelligence Unit (EIU) on the corresponding country. EIU provides regular reports on 150 countries. These reports describe the historical and current political trends, as well as macro- and sectoral economic activities. EIU also provides an electronic database from which economic statistics may be downloaded and manipulated. The electronic interface includes a tool to analyze country risk.

Another major source of country information is through the regular publications of the IMF, the World Bank, and the regional development banks. The IMF publishes the series *World Economic Outlook,* which analyzes the global economic trends, with a separate discussion of developing countries. It also publishes the series *Regional Economic Outlook,* which is composed of separate reports for each world region; these reports review the economic performance of and economic issues in the countries of each region.

The World Bank's most relevant regular publications are *Global Economic Prospects, World Development Indicators,* and *Global Development Finance* (formerly *World Debt Tables*). *Global Economic Prospects* contains an outlook on growth, inflation, interest rates, and so on. *World Development Indicators* contains numerous data series for monitoring progress in about 150 countries. "Progress" is considered to be multifaceted and encompasses economy, markets, people, environment, and so on. *Global Development Finance* is now recognized as an official source for studying countries' external debt and flow of funds. In addition, the World Bank has numerous other country reports that are not periodicals but, when available, provide comprehensive analysis of economic conditions in the corresponding countries. These reports are listed on the Web site of the World Bank and are available free of charge or at nominal prices.

Regional development banks have periodicals and ad hoc publications on the countries in their jurisdiction. A well-known annual publication by EBRD, *Transition Report,* provides indicators on economic reform. Many ad hoc publications provide in-depth analysis of various country issues. Examples are books by the IDB—for example, *Living with Debt, From Growth to Prosperity,* and *Revitalizing the Jamaican Economy.*

Country energy demand and supply

The eventual source of energy demand/supply information is through the relevant entities in the country (e.g., the electricity company, the oil and gas company, and the ministry of energy). However, convenient outside sources exist that often provide very valuable background for preparation of the analysis.

Information regarding a country's energy demand/supply, as well as policy issues and emerging trends, can be found in publications by and Web sites of the U.S. Energy Administration Information (EIA), the International Energy Agency (IEA), the World Bank and regional development banks, the United Nations, and industry sources such as BP, Cedigaz, and *Oil and Gas Journal.* A useful starting point is EIA's *Country Analysis Brief,* which is available on EIA's Web site for most developing countries. The second-most-immediate source is IEA, which publishes energy statistics and balances of OECD and non-OECD countries. IEA also publishes regional energy reports with reasonable coverage of developments and issues in each country.

The World Bank and other multilateral agencies publish numerous reports on energy sector conditions in developing countries. However, these reports are not published at regular intervals. Publications that are especially helpful for preparation of energy projects include

- Project appraisal documents (PADs), prepared by the World Bank and regional development banks for each loan made to a country. The reports on oil, gas, and power loans contain analysis of the energy sector.
- Energy sector studies, conducted occasionally by multilateral institutions, include studies of the oil, gas, and power sectors. These studies provide detailed analysis of energy demand/supply, structural and policy issues, and forthcoming developments.
- ESMAP reports. ESMAP is managed by the World Bank and carries out strategic studies of energy issues and options in developing countries. It has a large number of reports that are focused on a specific theme (e.g., renewable energy) or a specific country.

In addition to the aforementioned sources of information and analysis, the following sources provide data and statistics that may be useful to analyze the trends:

- The United Nations publication *Energy Statistics Yearbook* includes time series on energy production, consumption, and trade for all fuels and all countries.
- Another United Nations publication, *Energy Balances and Energy Profiles,* includes a country-by-country energy balance, energy conversion, and electricity production.
- The Cedigaz publication *Natural Gas Year in Review* is an annual survey containing time-series data on volumes of gas production, flaring, venting, reinjection, export, and import. It also discusses trends in gas use and prices in major gas-consuming countries.

- A PennWell publication, *International Petroleum Encyclopedia,* includes extensive maps listing locations of oil and gas reserves, refineries, terminals and transmission pipelines. It also contains reports on recent developments in upstream and downstream activities.
- The *BP Statistical Review of World Energy* concisely presents statistical trends in production, consumption, and international trade of coal, oil, and natural gas.
- Several journals, including *Oil and Gas Journal, Power Engineering, Independent Energy, Independent Power,* and the *Petroleum Economist,* provide coverage of energy issues and prospects in developing countries.

Global energy outlook

A number of institutions, including most major oil companies, analyze the global energy outlook. The following two sources provide long-term forecasts of demand/supply and crude oil prices on a regular basis and make their results available to the public:

- *World Energy Outlook,* published annually by IEA, contains forecasts of world energy demand/supply and oil prices, under various assumptions. Worldwide and regional forecasts are presented.
- *International Energy Outlook,* published annually by EIA, contains forecasts of world energy demand/supply and prices by type of fuel (coal, petroleum, or gas). It also includes an analysis of the impact of energy prices and technology on the development and growth of the various types of energy. Worldwide and regional forecasts are presented.

Summary 8–1. Helpful Sources of Information for Preparation of Energy Projects

International energy trends such as price of crude oil, world energy demand and supply, and so on, can be found in

- IEA's *World Energy Outlook*
- EIA's *International Energy Outlook*

Country-specific political and economic analysis and outlook, as well as risk analysis, can be found in

- EIU's country profiles and online risk analysis
- The IMF's *World Economic Outlook* and *Regional Economic Outlook*
- The World Bank's *World Development Indicators* and *Global Development Finance*
- Country-specific reports from the IMF, the World Bank, and regional development banks

Country-specific energy sector information and analysis can be found in

- EIA's country profile briefs
- IEA's regional and country-specific reports
- Project appraisal reports and policy papers by the World Bank and regional development banks
- ESMAP's thematic and country-specific reports
- The United Nations' *Energy Statistics Yearbook* and *Energy Balance*
- Industry publications by Cedigaz and BP, as well as *Oil and Gas Journal* and other periodicals

9 Establishing Economic And Financial Viability

Economic and financial viability are the priority concerns of investors and financiers examining a project. These factors must therefore receive priority attention in the project package. Indeed, analysis of financial and economic costs and benefits is the core of a project document, because it brings the results of all other analysis into an all-encompassing framework and indicates the risk-reward profile and, therefore, the attractiveness of a project.

Computation of economic and financial ratios is based on streams of capital costs, O&M costs, and project benefits (e.g., revenue from selling the output). The economic and financial analysis is concerned with

- Review of all factors that affect capital costs, O&M costs, and project revenues and other benefits, including quantification of the input of all these factors and arrival at one set of figures that represent the most likely streams of costs and benefits
- Calculation of relevant economic and financial ratios
- Comprehensive risk analysis based on possible variations in cost and benefit streams

Economic versus Financial Analysis

Economic analysis usually refers to assessment of the costs and benefits of a project to the society or the country as a whole. *Financial analysis* refers to assessment of the costs and benefits of a project to the company undertaking the project. For example, if electricity consumers receive a benefit equivalent to 8¢/kilowatt-hour (kWh) from using electricity but pay only 5¢/kWh, the economic analysis would use 8¢/kWh as an indication of benefits, whereas the financial analysis would use 5¢/kWh because this is the actual revenue from selling the electricity.

The distinction between economic analysis and financial analysis was most pronounced during the 1960s and 1970s, when prices were more controlled and distortions were greater. During this period, substantial theoretical work was done to derive methods of quantifying economic costs and benefits. These methods adjust the observable cost and benefit figures to arrive at the corresponding economic values.

Most observable figures are financial costs and revenues. Examples are the actual price received by a utility for selling electricity, the actual exchange rate at which the company converts foreign currency, and the actual wage paid to the laborer. These figures, however, may not indicate the true costs and benefits to the economy. First, taxes, royalties, economic rents, and so on, are all considered as cost to the company but are indeed money received by the government; therefore, these components do not represent a true cost to the economy. Second, market distortions caused by controls, subsidies, and other market failures result in prices, exchange rates, and interest rates that do not correspond with competitive market equilibria.

Adjustments of the observable cost and benefit figures to reflect taxes and royalties are generally straightforward and are routinely done in project analysis. The adjustments to compensate for market distortions are more complicated and normally involve estimating shadow prices, shadow exchange rates, shadow wage rates, and so on, which reflect the economic values of these parameters. This latter type of adjustment became less important in the 1980s and 1990s, as most developing countries moved toward market economic systems and as energy prices and exchange rates became more market related.

For many energy projects, an additional adjustment should be made to take account of a *depletion premium* if the proposed plant produces or uses exhaustible resources. Again, in an economically efficient environment, financial prices incorporate this factor.

Economic and financial analysis should start with financial costs and benefits—that is, the project's expenses and revenues. These figures form the basis of financial analysis. In contrast, economic analysis should

- Deduct from project costs those funds paid to the government for royalty, production tax, asset tax, import tariffs, and so on. These items, although representing costs to the company, are not a cost to the economy. Rather, they are simple transfers of funds from the company to the government.
- Add to project revenues the sales tax collected from consumers. Again, this amount does not represent a benefit to the company but is part of the overall project benefits (i.e., to the economy as a whole). In other words, the revenue from the project includes all the money paid by consumers, even while part of this revenue is transferred to the government and the rest remains with the company.

INCREMENTAL VERSUS TOTAL COSTS AND BENEFITS

The analysis of costs and benefits of a project should be based on the project's corresponding costs and revenues. In certain investments—for example, construction of a new power plant or a grassroots refinery—the project is an independent facility with clear and measurable output. In other investments—for example, expansion of power or gas distribution systems—the project adds components to an existing facility. The primary attempt is normally aimed at assessing and analyzing *incremental* costs and benefits—that is, those costs and benefits that are associated with the proposed investment. When this is not possible, the analysis is based on total facilities.

METHODS AND RATIOS

Two key measures used by investors to assess the economic and financial viability of a project are the *net present value* (NPV) and the *internal rate of return* (IRR) of the project (fig. 9–1). The NPV of a project is the discounted value of cash inflows minus the cash outflows of the project. To calculate the project NPV, a value has to be assigned to the discount rate. This value, a market interest rate relevant to the project, indicates the cost of money. Thus, the NPV measures the actualized net revenue (i.e., the revenue minus all the costs) and indicates the attractiveness of an investment to the sponsors. In other words, the NPV indicates the difference between an investment's market value and its costs—and, thus, the reward for sponsors for putting capital at risk to undertake the project. For example, an NPV (at a 10% discount rate) of $20 million indicates not only that all the initial project expenditures, including the capital funds put in by sponsors, earn a 10% annual return but also that there is an additional net revenue whose present value is $20 million.

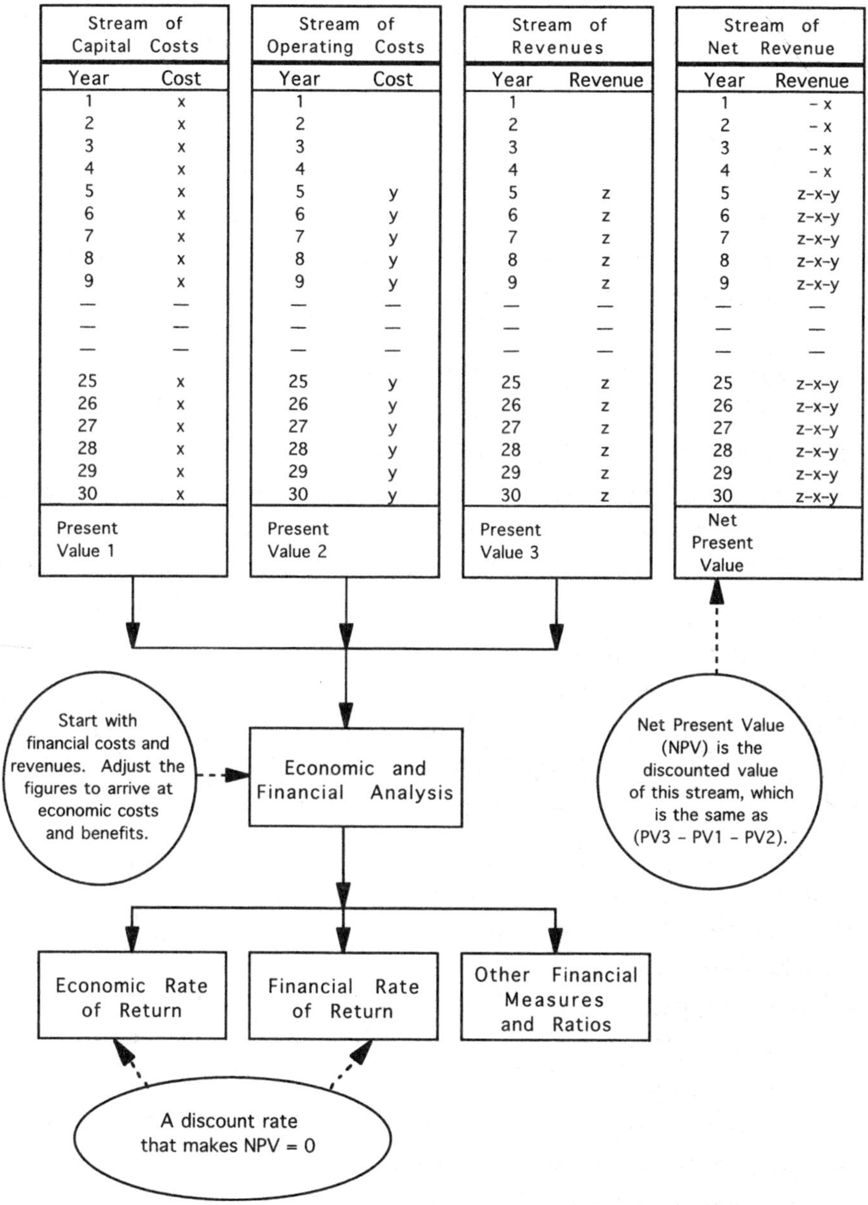

Fig. 9–1. Calculation of NPV and IRR

The IRR is the discount rate that makes the NPV of a project zero. An NPV of zero indicates that the project will earn a rate of return equal to the discount rate.

Lenders' return on a project (i.e., interest received on the loan) is determined in advance and does not change if the project turns out to be more profitable or less successful than initially appraised. Lenders have legal assurances, based on well-designed security packages, that principal and interest on the loan will be paid. Nevertheless, they examine carefully the NPV and the IRR to make sure that project sponsors have a strong financial incentive to undertake the project and see it through to completion.

The IRR is further analyzed to derive the appropriate numbers for return on equity investments. The IRR measures return on the project as a whole. If the project is funded through 30% equity and 70% debt, the project's IRR will approximately follow the formula

Project IRR = 0.3 (Return on equity) + 0.7 (Average interest rate on debt)

For example, if the project IRR is 15% and the average interest on the debt is 10%, then the return on equity will be about 27.

The obtainment of a desirable IRR depends very much on project risks and on whether the project is undertaken by private or state sponsors. Private sponsors want at least 25%–35% return on equity. This translates into a project IRR of more than 15%. State sponsors normally require about 10% return on equity, which would translate into a 10% project IRR, assuming that the interest rate is also 10%.

State-sponsored investments emphasize *economic IRR*. As a general guideline, the economic IRR of the project should not be less than the country's overall opportunity cost of capital, a discount rate normally assessed by the main economic body of the country (e.g., the planning agency or the ministry of finance).

Lenders derive comfort from sponsor's equity investment in the project at a level commensurate with risk. For example, power plants support higher *gearing,* or debt versus equity ratio, than cyclical refineries or petrochemical plants. However, in limited-recourse or nonrecourse project financing, the value of the asset is tied to the cash flow that can be generated and made available for debt service—that is, scheduled interest and principal payments. Thus, lenders place more emphasis on projected cash flow analysis than on return on investment.

Several ratios are calculated on the basis of the project cash flow. For example,

- DSCR, which reveals the cash flow available for debt service, calculated for each year on a cumulative basis
- Loan-life coverage ratio, which divides the outstanding debt into the remaining cash flow stream
- Interest coverage ratio, which is the ratio of available cash flow over interest payments

The lenders' covenants set certain floor values for these ratios. The floor value, always greater than 1.0 (e.g., 1.3), varies depending on the type of project and financing structure. The floor value may be revisited at project completion when recourse to project sponsors is due to be reduced substantially or eliminated.

If coverage ratios fall below the specified floor levels, lenders have the right to call the loan and exercise their lien over the project, after appropriate notice. However, in practice, such situations are usually avoided through preventative measures. For example, lenders and sponsors normally set up a disbursement account, with a trustee, that receives all project revenues in escrow and pays out funds according to a defined order of priority. Whenever risks regarding foreign exchange availability or convertibility are present, an offshore disbursement account is established to pay foreign exchange obligations, including interest and principal payments of foreign loans.

Upstream Oil Projects

Economic and financial analysis of upstream petroleum projects involves a number of special issues (fig. 9–2). The most important of these issues are the choice of an acceptable price forecast for crude oil, the appropriate accounting of the size of oil reserves and their producibility, the incorporation into calculations of the costs and benefits of any associated gas (or liquefied petroleum gas [LPG]), and the consideration of taxes, royalties, and "profit oil."

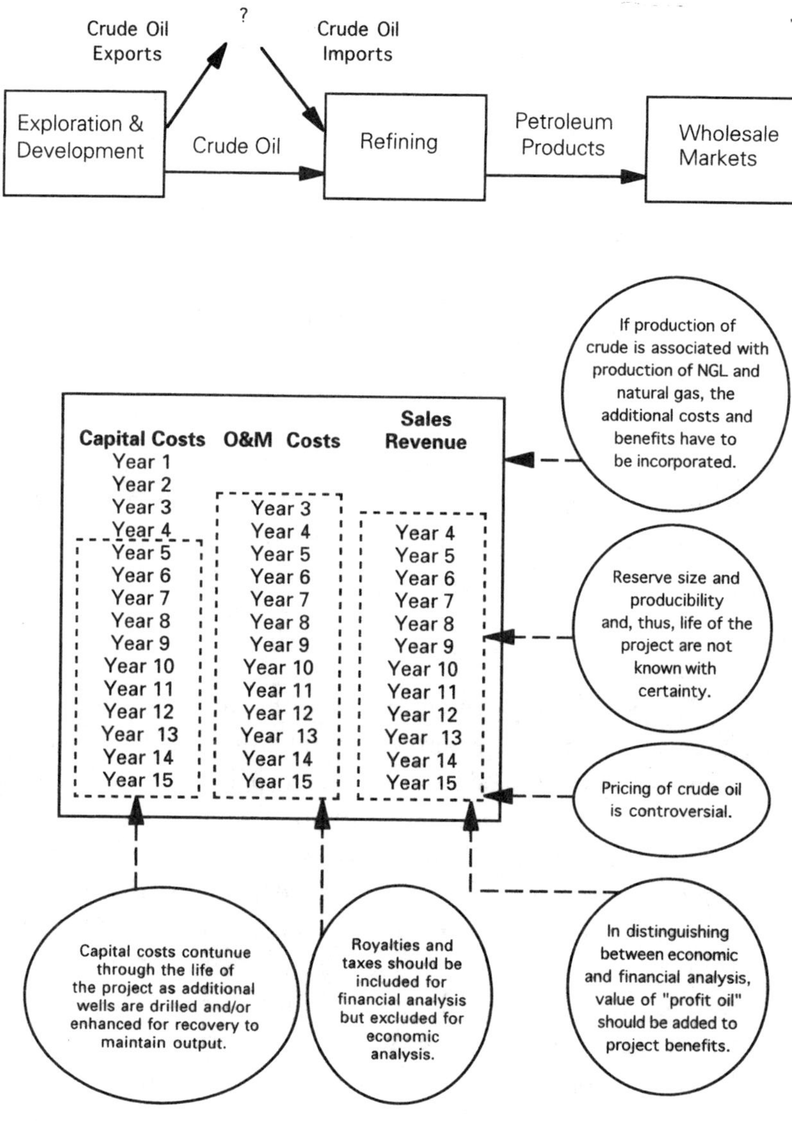

Fig. 9–2. Petroleum exploration and development projects: Special issues in economic and financial analysis

Pricing of Crude Oil

The price of crude oil represents the most important piece of information for an upstream petroleum project. However, pricing of crude oil is also a very controversial matter. The controversy emerges from two sources:

- Differences among various forecasts of the international price
- The way in which the international price becomes relevant to the project

The first type of controversy has now become part of the petroleum business, and investors and financiers have learned to live with it. It is accepted practice to take a reputable published forecast as the base-case scenario and carry out sensitivity analysis to quantify the risks associated with lower prices. For long-term forecasting of crude oil, many analysts use the price projections of IEA and EIA.

The second type of controversy, the manner in which the international price becomes relevant to the project, must be addressed and resolved before the final application for financing. If all of the project output is exported, then the matter is rather straightforward, and *export parity,* the price minus the associated transportation cost, is used as the basis for pricing the crude. However, in most oil projects, the output is either fully or partially sold to domestic refineries. In such a case, planners must estimate the price at which the producer can sell the crude. Financiers normally want to see that the producer has a firm contract with a clear price, or price formula, with domestic refineries.

The price negotiated between a producer and a refiner depends on the overall demand/supply picture and alternative options for either side. In the event that the proposed upstream project is the only source of domestic crude supply, the producer may force the refiners to pay a price close to *import parity,* the international price plus cost of transportation. In the event that project output represents one source of crude among many, the refiner likely would not pay more than the export-parity price.

Economic pricing of crude oil—that is, the pricing used in economic, rather than financial, analysis—is determined in the following manner:

- If the country is short of crude oil and relies on imports, output of the proposed project will replace foreign oil. The benefit to the economy is estimated on the basis of the import-parity price of crude oil.
- If the country has a surplus of crude oil, output of the proposed project would be exported or would replace that of some other producers, who would be forced to export. The benefit of the project to the economy is estimated based on the export-parity price of crude oil.

The Relevance of Netback Value

Wherever the pricing of crude oil is subject to question, the seller and buyer consider the oil's *netback value*. This is an estimate of the value of products that a given quantity of specific crude will yield after processing by a particular refinery, based on product prices in a particular market. Specifying the crude type and refinery characteristics is important, because variations in either would change product yield patterns central to the calculation.

Steps involved in the calculation of netback value depend on the stages of physical movement of crude oil from the wellhead to the refinery. In a typical case, the physical movement of oil is from field to loading terminal, from loading port to receiving port via tanker, from receiving terminal to refinery, and from refinery to market. Computation of netback value follows these same movements but in reverse order (fig. 9–3), as follows:

- First, the weighted average value of the refined product obtainable from a barrel of crude oil at the refinery gate is computed. The result is known as the *gross product worth* (GPW) of the crude. The GPW is calculated by multiplying the prevailing price for each product by its percentage yield from one barrel of crude oil.
- Then, the cost of refining, which consists of out-of-pocket operating expenses involved in handling the last barrel of crude by a refiner, is deducted from the GPW to arrive at the net product worth (NPW) of the crude. Since the marginal barrel is under consideration, this result does not include any amortization or depreciation.
- Finally, the costs of transportation and insurance are deducted from the NPW to arrive at the netback value of crude at the port of loading.

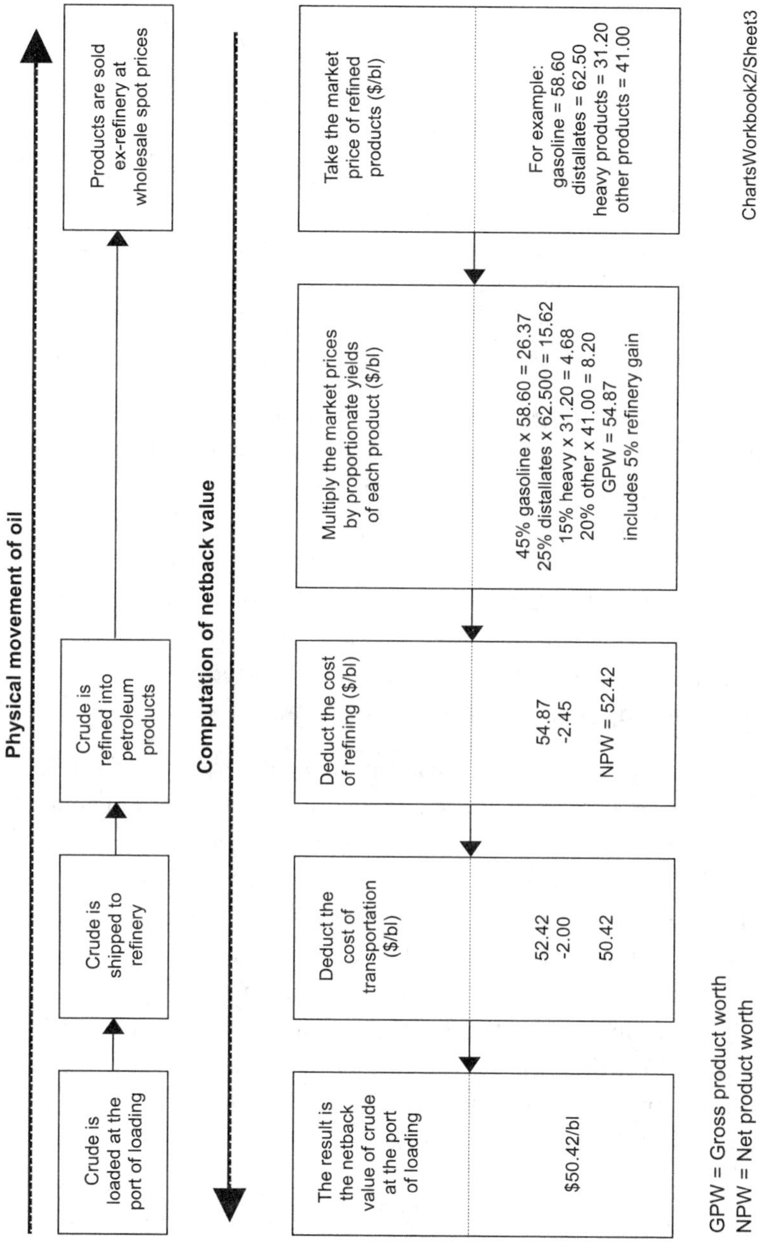

Fig. 9–3. Calculation of netback value of crude oil

The most important information in this calculation is the *refinery yield*—that is, the mix of products obtained from a barrel of crude oil. This mix depends not only on the quality of the crude but also on local demand patterns and the technical capabilities of individual refineries. Thus, looking for specific yield patterns would require compiling thousands of mixes corresponding to various crudes, various refineries, and different periods. Recognizing the impossibility of such a task, *Petroleum Intelligence Weekly* has developed data on crude oil yield patterns that are typical or representative of the refining industry in each of the six major refining centers.

Further, the refining cost, used in calculations of netback value, does not include capital cost. The implication is that netback value is purely a marginal phenomenon that corresponds with the short-term operation of a refinery. Thus, netback value, as calculated and published in industry papers, is not intended as a basis for long-term development and resource management. However, a combination of netback value and capital cost of refining may be used in long-term analysis when other measures are not available.

Netback value is often used in negotiations of the crude price supplied to captive markets, in which petroleum product prices are not linked to international prices. Refineries in these markets charge product prices substantially different from international prices, and this is often the case in developing countries. Thus, the value of crude oil supplied to these refineries may differ from international crude oil prices. In such situations, the netback value of the specific crude oil is calculated on the basis of the refinery yield and prices in the domestic market. The netback value will then become a basis for negotiating the crude oil price.

Size of oil reserves

For upstream projects, financing institutions make sure that sufficient oil reserves are present to support projected production profile and, therefore, repayment of the debt. This is usually done by obtaining independent certification of the reserves by an internationally reputable company. Reserves are classified into three categories:

- *Proven reserves,* which can be recovered with a high degree of certainty
- *Probable reserves,* which are likely to be proven but are included in part of the field that requires further delineation or need confirmation of higher recovery by enhanced recovery methods
- *Possible reserves,* which involve high risk with regard to recovery

Conservative financiers consider only proven reserves. However, in the early stages of field development, it is common to include risk-weighted probable reserves, to estimate production profiles and field life. The risk factor applied to these reserves varies between 30% and 50%, depending on the degree of confidence in the estimate of the probable reserves.

Other issues

Economic and financial analysis of upstream oil projects should also consider the following:

- Capital expenditures normally continue during the life of the project to maintain or increase output, for example, by the drilling more wells or implementing enhanced recovery systems, and, finally, for decommissioning.

- In certain upstream projects, part of the output is transferred to the government (usually the NOC) under terms of a production-sharing contract. The value of this amount of oil is not included in project benefits when carrying out financial analysis. However, the value of this output should be incorporated into project benefits when computing economic rate of return.
- The royalty and the production tax paid to the federal and local governments are included in O&M costs for financial analysis. They should be excluded from O&M costs in calculations of economic rate of return.
- Often an oil field produces natural gas liquids (NGL), in addition to crude oil. The NGL output should also be priced appropriately and added to project benefit.
- In certain fields, crude oil is produced in association with natural gas. The associated gas is flared, reinjected, or collected for consumption. In each situation, the costs and benefits have to be incorporated in the analysis (for further details, see the next section, "Upstream Gas Projects").

Upstream Gas Projects

Economic and financial analyses of gas production must address issues similar to those of upstream oil projects, as discussed in the previous section. However, an important parameter for analysis—the price of natural gas—is determined in a different manner from that of crude oil.

Gas pricing

Gas pricing is a complex issue because

- Governments often intervene to some degree in determining gas prices. In such cases, the issue, even for government authorities themselves, is to find a price level (or price formula) that is fair to consumers and producers and that results in economical production and consumption of gas. Quite often, social and political considerations make the price setting more complex.
- Negotiations between sellers and buyers are based on a host of variables, some of which are subjective and difficult to quantify. This complexity is also highly dependent on the industry structure and the shares of the public and private sectors in the industry ownership.
- No international price exists for natural gas, although the spot LNG and forward gas markets are used for reference.

A host of factors affects the views of sellers and buyers about the price of gas. For the seller, the price should cover the cost of production and (if applicable) the cost of processing and transportation. In addition, the seller will attempt to receive a higher price to cover the intrinsic value (also referred to as *economic rent* or the *depletion premium*) of natural gas. Further, every seller wants to maximize profits—or at least to include a just reward for exploration risks. Since a "normal" return is included in the cost of supply, the additional profits are considered to be "above normal."

For the buyer of gas, the starting point for assessment of price is the cost of replaced fuel. If gas is replacing coal, then the price of coal, after certain adjustments to take into account differentials in capital cost and efficiency, represents the value that the buyer receives from using natural gas. In addition, the user will receive benefits such as environmental advantages, ease of use, and security of supply. Nevertheless, the buyer does not want to pay for all these benefits. Indeed, the buyer normally argues for a price low enough to provide cost-saving incentives to switch to gas.

Although the aforementioned considerations establish a conceptual framework for discussion of the gas price, calculation of a price figure is not a straightforward matter, for the following reasons:

- Several of the variables (e.g., economic rent and environmental benefits) are not easily quantifiable
- The seller wants to charge a price higher than indicated by the costs, and the buyer wants to pay a price lower than indicated by the benefit of using gas

Thus, in most price negotiations, the seller refers to the benefits that the buyer receives from using gas to justify a higher price, whereas the buyer refers to cost of supply to limit the price to out-of-pocket supply costs.

Hence, price negotiations may result in any price between the cost of production and the economic value of gas. The negotiating positions of gas sellers and buyers depend on the structure and organization of the gas sector. In some developing countries, the gas sector is owned and operated by an integrated national gas company. In such cases, the gas price is set by the government (or national gas company), and negotiations take place with large consumers, such as power companies and industrial customers.

In other countries, upstream investments are undertaken by the private sector. The gas price is either announced by the government or negotiated between the producer and the national gas company. In the few countries where transmission and distribution (T&D) activities are in the private sector, producers negotiate gas prices with distribution companies and, sometimes, directly with major consumers.

Economic framework for gas pricing

In most developing countries, the government plays a direct or an indirect role in setting gas prices. Governments normally take an economic (rather than financial) perspective, to achieve a socially optimum solution. The economic framework (fig. 9–4) indicates that the gas price should be set somewhere between the long-run marginal cost (LRMC) and the netback value, depending on the country's endowment of gas reserves. LRMC measures the cost of gas supply. The addition of depletion premium to LRMC results in economic cost of supply. The netback value measures the benefit to the consumers of using natural gas; it indicates the upper limit of the price range, because if the price were set above the netback value, the consumer would switch to an alternative fuel.

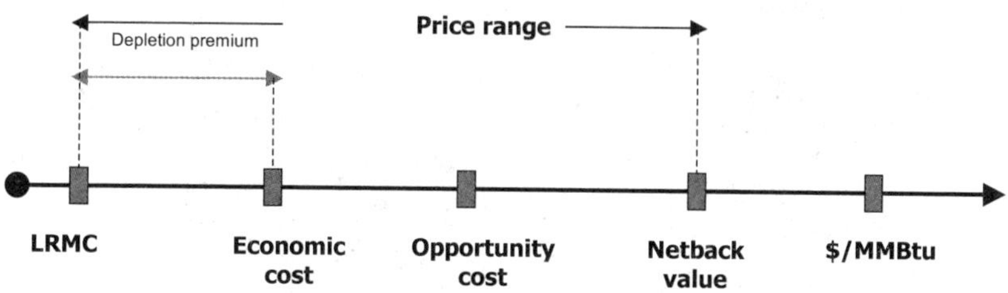

Fig. 9–4. Economic framework for gas pricing

Finally, a measure that has recently become relevant is the net gain from gas exports, particularly in the form of LNG. This is normally referred to as the *opportunity cost,* to imply that the seller could export the gas instead of selling it to the domestic market.

THE RELEVANCE OF NETBACK VALUE

The concept of netback value is often used in price negotiations and by governments, to make decisions regarding taxes and other gas-related policies.

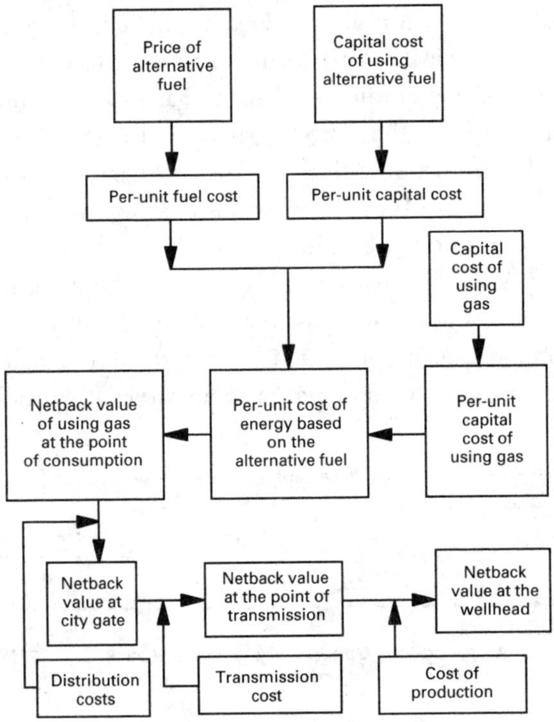

For financial netback value, market prices of alternative fuels and capital equipment are used.

For economic netback value, market prices should be adjusted to exclude taxes and duties.

Fig. 9–5. Calculation of netback value of natural gas

For each gas supply system, the netback value can be calculated at various points of delivery (fig. 9–5). The starting point for calculation of the netback value is the ending point in the delivery system. In other words, the netback value is first calculated for the point of final consumption (at the burner tip), based on capital and fuel costs of the alternative fuel.

For example, if gas is replacing coal in power generation, the costs of generating power from a coal-based plant should be compared with those of generating power from a gas-based plant (table 9–1). The netback value for gas is the gas price at which the final costs per kilowatt-hour would be the same for coal-based and gas-based electricity. Thus, the calculation procedure involves

- Estimating the costs per kilowatt-hour of fuel and capital for a coal-fired power plant. The sum of these (fuel and capital) costs gives the overall cost per kilowatt-hour of electricity generation by the coal power plant.
- Estimating the per-unit capital cost of the gas-based power plant.
- Deducting the second of these from the first—that is, assigning a per-kilowatt-hour fuel cost to gas that would make the overall per-kilowatt-hour cost of gas-based power generation the same as that of coal-based generation.
- Converting the per-kilowatt-hour fuel cost to the equivalent price per million British thermal units (MMBTU) or price per thousand cubic feet (MCF) for natural gas, based on the thermal efficiency of the plant and the heat content of the gas. The resultant figure is the netback value of gas at the point of consumption.
- Netback value can be calculated for other points in the delivery system, such as at the wellhead or ex-transmission, by deducting the appropriate costs.

TABLE 9–1. CALCULATION OF NETBACK VALUE OF GAS USED IN POWER GENERATION

	Coal-Fired Steam Plant	Gas-Fired Combined-Cycle Plant
Generating Capacity	300 MW	300 MW
Load Factor	75%	75%
Output	1971 GWh/y	1971 GWh/y
Thermal Efficiency	38%	49%
Capital Cost	US$1250/kW	US$600/kW
Plant Cost	US$375 million	US$180 million
Life of the Plant	30 years	30 years
Per Unit Capital Cost	2.2 ¢/kWh	1.1 ¢/kWh
Fuel Price	$2.10/MMBTU	$4.70/MMBTU
Per Unit Fuel Price	1.9 ¢/kWh	3.2 ¢/kWh
Other O&M Costs	0.5 ¢/kWh	0.3 ¢/kWh
Generation Costs	4.6 ¢/kWh	4.6 ¢/kWh

Notes: Per-kWh generation of coal-based power is calculated from the data and equals 4.6 ¢/kWh.

Per-kWh capital and O&M costs of gas-based power are calculated from the plant data and equal 1.1 ¢ and 0.3 ¢/kWh, respectively.

The fuel price for gas (3.2 ¢/kWh or $4.70/MMBTU) is calculated so that the overall cost of gas-based power is the same as that of coal-based electricity.

Netback value is normally calculated for alternative uses of gas—for example, gas used in the residential sector or in the industrial sector (in the form of fuel or petrochemical feedstock). The calculation results in a range of netback values. Theoretically, the economic benefit of additional (marginal) gas supply is indicated by the lowest netback value among the sectors that consume gas. However, in the discussion of gas prices, the negotiating parties review the entire range of netback values.

The relevance of the depletion premium

Like oil, natural gas is often viewed as an exhaustible resource. For this reason, the economic cost of gas supply consists of two components:

- The cost of gas production (and T&D if relevant)
- A depletion premium, indicating the opportunity cost of consuming a unit of the depletable resource now, rather than in the future

Assessment of the depletion premium for natural gas involves the following steps:

- Projecting gas demand and determining the year (switching time) in which further growth in gas demand would not be sustained by the proven gas reserves. The limited availability of gas would then force the country to satisfy incremental demand with an alternative fuel.
- Projecting the price of the alternative fuel to the switching time and adjusting this price to take into account differentials in capital and O&M costs and thermal efficiency.

- Calculating the difference between the adjusted price (obtained in the preceding step) and the cost of gas production at the switching time. This difference represents the depletion premium at the switching time.
- Computing the depletion premium for prior years by appropriately discounting the premium at the switching time.

As indicated by the preceding methodology, the depletion premium would be small in gas-rich (gas-surplus) countries. For a gas-poor country, the switching point is the present, which means that the calculation of the depletion premium coincides with the calculation of the netback value. Therefore, the concept of the depletion premium provides a general framework that can be used to draw the following conclusions:

- In countries with abundant gas reserves, the depletion premium is negligible. The cost of gas supply is approximately the cost of production. The gas price is set either by the government, based on cost considerations, or by the free market, which also results in prices close to cost of supply. The resultant price could be substantially lower than the price of the alternative fuel.
- In countries with shortages of gas, economic analysis should be based on netback values. Also, in a practical sense, gas prices are either set by the government, at levels close to the netback value, or determined in a free market that pushes price toward the netback value.

- In countries that have sufficient gas only for a limited number of years (referred to as the *surplus window*), the price of gas should not be as high as the netback value but should be higher than the cost of production. In such cases, a depletion premium should be calculated and added to the cost of production to arrive at the economic cost of gas supply.

Other Issues

During economic and financial analysis of upstream gas projects, a number of other issues may arise:

- The produced gas may be associated with liquids (e.g., condensates and LPG). The costs and benefits should be adjusted if the liquids are separated and sold.
- The produced gas may include a significant amount of ethane. The ethane can be separated and sold as feedstock to petrochemical plants. In such a situation, costs and benefits should be adjusted accordingly.
- In calculations of netback values, components of the price of the alternative fuel should be examined carefully. In most developing countries, fuel prices include a variety of taxes. Whereas the financial analysis is based on out-of-pocket cost to the fuel consumer, the economic analysis is based on the cost to the economy and therefore should use prices that exclude taxes and duties (table 9–2).
- In situations where gas is used in several sectors (e.g., power, industrial, and residential), separate netback values have to be calculated for each type of use. Economic analysis should weight each netback value by the amount of gas consumed in the corresponding use.

TABLE 9–2. ECONOMIC AND FINANCIAL ANALYSIS OF UPSTREAM GAS PROJECTS

	Financial Costs ($ million)			Financial Benefits ($ million)			
Year	Capital	O&M	Total	Production (BCF)	Price ($/MCF)	Revenues ($ million)	Net Benefits Stream ($ million)
1	30.00	6.00	36.00	0.00	0.0	0.00	-36.00
2	230.00	20.00	250.00	0.00	0.0	0.00	-250.00
3	360.00	60.00	420.00	0.00	0.0	0.00	-420.00
4	240.00	100.00	340.00	0.00	0.0	0.00	-340.00
5	70.00	100.00	170.00	85.00	4.2	357.00	187.00
6	120.00	100.00	220.00	90.00	4.2	378.00	158.00
7	96.00	100.00	196.00	110.00	4.2	462.00	266.00
8	28.00	100.00	128.00	105.00	4.2	441.00	313.00
9	50.00	104.00	154.00	95.00	4.2	399.00	245.00
10	80.00	104.00	184.00	95.00	4.2	399.00	215.00
11	64.00	112.00	176.00	95.00	4.2	399.00	223.00
12	60.00	104.00	164.00	95.00	4.2	399.00	235.00
13	60.00	84.00	144.00	95.00	4.2	399.00	255.00
14	0.00	80.00	80.00	95.00	4.2	399.00	319.00
15	0.00	80.00	80.00	95.00	4.2	399.00	319.00
16	0.00	60.00	60.00	95.00	4.2	399.00	339.00
17	0.00	50.00	50.00	95.00	4.2	399.00	349.00
18	0.00	60.00	60.00	95.00	4.2	399.00	339.00
19	0.00	60.00	60.00	95.00	4.2	399.00	339.00
20	0.00	60.00	60.00	95.00	4.2	399.00	339.00
PV @ 10%	**936.26**	**625.49**	**1561.74**	**510.98**		**2146.13**	**584.38**

Financial IRR = 18% **Economic IRR = 24%**

Notes: For financial analysis, the capital and O&M costs are based on projected expenses of the company and include taxes and duties. The gas price is the net price to be received by the company.

For economic analysis, capital and O&M costs exclude taxes and duties (estimated at 5% of total cost). The gas price (estimated at $4.85/MCF) is based on the netback value of gas. The choice of gas price for economic analysis would depend on whether the country has a surplus or a shortage of gas. This case is adopted from a project in Thailand, where the country has a shortage of gas, and therefore, netback value is an appropriate measure of economic benefit of using gas.

PV = Present Value
BCF = billion cubic feet
MCF = thousand cubic feet

Gas T&D

Economic and financial analyses of gas T&D projects must answer three important questions: First, can the new investment be considered as a separate facility, or would it be intertwined with the existing system? Second, can the T&D company own (buy and sell) gas, or would it function as only a carrier of gas? Third, can the economic rate of return be calculated satisfactorily?

Separate versus intertwined facilities

Clearly, economic and financial analysis of a T&D project is more straightforward when the proposed investment deals with independent facilities. This would be the case for projects that build new supply systems. For such projects, cost estimates are based on international norms for capital and O&M expenditures. Project benefits are assessed on the basis of the price margin or the throughput tariff of the proposed system.

A new investment intertwined with existing facilities requires an incremental analysis. This involves examining all of the facilities and comparing the gas delivery system with and without the proposed investment. The difference could be only in the volume of the delivered gas or in the reliability of the system, maintenance costs, and so on. All the differences have to be identified, estimated, and incorporated in costs and benefits of the new investment.

Ownership versus carrier function

In most developing countries, T&D companies are permitted to own the gas that they carry. They buy the gas from producers and deliver and sell it to consumers. Thus, in the analysis of project costs and benefits, the cost of purchasing the gas is included in the operations costs. The revenue from selling the gas constitutes the project benefits. If the company is selling to various customer groups, then the average tariff for each group and the relative volumes of gas supplied to each group are used in assessing the project benefits.

In countries with significant private sector participation in the gas industry, transmission companies may be barred from owning the gas that they carry. Instead, as *common carriers,* they are required to carry the gas for others at a regulated tariff. This arrangement is meant to prevent transmission companies from exercising monopoly power in the gas market. The throughput tariff covers capital and O&M costs of the pipeline company. The cost-benefit analysis is relatively straightforward, because project revenues can be calculated simply on the basis of throughput volumes and tariffs.

Calculation of Economic Rate of Return

For most T&D projects, estimates of a financial rate of return can be based on T&D costs and revenues (table 9–3). However, calculation of an economic rate of return is not feasible unless the entire gas chain is considered. Thus, economic analyses often view investments in gas production and T&D as an integrated undertaking.

Separate estimation of the economic return for the T&D project is not possible because the benefit due to T&D cannot be isolated from the total benefit. For example, if the cost of gas production is $1.00/MMBTU and cost of T&D is $0.80/MMBTU, then the total cost is $1.80/MMBTU. If the consumer receives a benefit of $5.00/MMBTU from using this gas, then the net economic benefit is $3.20/MMBTU. However, there is no way to say how much of this benefit is related to T&D and how much is related to production and development. In effect, the benefit is brought about by the whole chain.

TABLE 9–3. FINANCIAL ANALYSIS OF A GAS TRANSMISSION PROJECT

Year	Capital cost ($ million)	O&M cost ($ million)	Volume of gas (BCF)	Purchase price of gas ($/MCF)	Cost of gas purchase ($ million)	Sale price of gas ($/MCF)	Sales revenue ($ million)	Net benefit ($ million)
1	88	0	0	0	0	0	0	-88
2	200	0	0	0	0	0	0	-200
3	75	12.5	60	4.20	252	5.00	300.00	-40
4	0	12.5	80	4.28	343	5.10	408.00	53
5	0	12.5	95	4.37	415	5.20	494.19	67
6	0	12.5	110	4.46	490	5.31	583.66	81
7	0	12.5	95	4.55	432	5.41	514.16	70
8	0	12.5	95	4.64	441	5.52	524.44	71
9	0	12.5	95	4.73	449	5.63	534.93	73
10	0	12.5	95	4.82	458	5.74	545.63	75
11	0	12.5	95	4.92	467	5.86	556.54	77
12	0	12.5	95	5.02	477	5.98	567.67	78
13	0	12.5	95	5.12	486	6.09	579.02	80
14	0	12.5	95	5.22	496	6.22	590.60	82
15	0	12.5	95	5.33	506	6.34	602.41	84
16	0	12.5	95	5.43	516	6.47	614.46	86
17	0	12.5	95	5.54	526	6.60	626.75	88
18	0	12.5	95	5.65	537	6.73	639.29	90
19	0	12.5	95	5.77	548	6.86	652.07	92
20	0	12.5	95	5.88	559	7.00	665.11	94
PV @ 10%	301.18	84.73	615.84		2946.46		3507.69	**175**
			IRR =	**17%**				

Note: The above analysis is based on a case involving construction of 175 km of 32-inch and 160 km of 24-inch pipelines to transport gas from upstream to the point of distribution (city gate).

BCF = billion cubic feet

An obvious exception is a country that imports natural gas. In such a case, the price of imported gas is a clear indication of the cost of gas to the economy. The difference between this cost and the netback value represents the benefit that the T&D project would provide by carrying the gas to consumers.

PETROLEUM REFINING

Economic and financial analysis of petroleum refining should consider that costs include investment and O&M costs, as well as the cost to purchase crude oil, and that revenues are based on the wholesale prices of petroleum products. The most important issue in the analysis is the forecasting of relevant crude and product prices. (For financial analysis, see fig. 9–6 and table 9–4.)

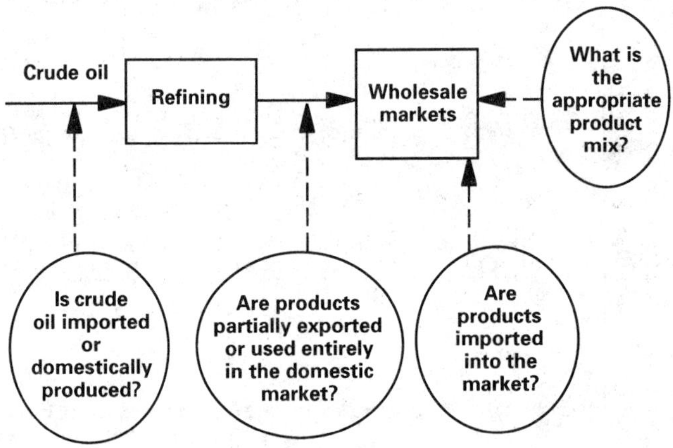

Fig. 9–6. Financial analysis of petroleum refining

Table 9–4. Financial analysis of a petroleum refining project

Year	Capital costs ($ million)	O&M costs* ($ million)	Volume of crude oil (b/d)	Price of crude oil ($/bl)	Cost of crude oil** ($ million)	Product Price ($/bl)	Product sales*** ($ million)	Net Revenue ($ million)
1	3.00	0.00	0.00	0.00	0.00	0.00	0.00	-3
2	52.50	0.00	0.00	0.00	0.00	0.00	0.00	-53
3	67.50	0.00	0.00	0.00	0.00	0.00	0.00	-68
4	63.00	6.00	5000.00	60.00	99.00	66.00	116.20	-52
5	0.00	14.00	15000.00	61.20	302.94	67.32	355.56	39
6	0.00	14.00	20000.00	62.42	412.00	68.67	483.56	58
7	0.00	14.00	25000.00	63.67	525.30	70.04	616.54	77
8	0.00	14.00	28000.00	64.95	600.10	71.44	704.34	90
9	0.00	14.00	28000.00	66.24	612.10	72.87	718.42	92
10	0.00	14.00	30000.00	67.57	668.94	74.33	785.14	102
11	0.00	14.00	30000.00	68.92	682.32	75.81	800.84	105
12	0.00	14.00	30000.00	70.30	695.97	77.33	816.85	107
13	0.00	14.00	30000.00	71.71	709.88	78.88	833.19	109
14	0.00	14.00	30000.00	73.14	724.08	80.45	849.86	112
15	0.00	14.00	30000.00	74.60	738.56	82.06	866.85	114
PV @ 10%	139.86	66.21	117195.93		2589.30		3039.07	244
			IRR =	**30%**				

* O&M costs include the cost of fuel used by the refinery.
** 330 days/year.
*** Based on the yield of various products (gasoline, jet fuel, gasoil, fuel oil, and butane) using a volume gain of 6.7%.

b/d = barrels per day
PV = Present Value

THE RELEVANT CRUDE OIL PRICE

The problem of forecasting the crude oil price is related to controversies about international price forecasts and possible distortions between domestic and international prices. If crude oil is imported, financial and, in particular, economic analyses are simplified; the economic price is based on import parity, and the financial price includes taxes and duties. However, if crude oil is produced domestically, the economic price assessment is based on export parity, but the financial price is difficult to forecast because it varies, depending on the market power of the refiner and the upstream producers; therefore, most financiers prefer to see that the refiner has a firm contract for crude oil supply if the source is domestic.

THE RELEVANT PRODUCT PRICES

Forecasting product prices is even more complicated than forecasting the crude oil price. First, there are no systematic forecasts for international product prices. The relationship between the crude oil price and product prices has been unstable over time and across countries and markets. Most attempts to formulate and forecast this relationship have not succeeded, although some approximations have emerged.

Second, in most developing countries, domestic product prices diverge from international spot prices of these products. Thus, the forecasting of product prices in a developing country requires an intimate understanding of the relevant laws, regulations, and government policy, as well as the demand/ supply patterns of the product market. Except for *telling* refineries, which operate on a cost-of-service basis, project operating margins will be highly cyclical. Cash shortfalls may affect debt service in cycle troughs; therefore, various mitigation measures may be necessary, such as high initial equity, shareholders' support, netback pricing of captive feedstocks, special reserves, and dividend clawbacks.

Other Issues

Economic and financial analysis of refining projects should give further consideration to the following issues:

- Forecasting of the products mix is an important determinant of project benefits. However, the products mix will depend on the configuration of the refinery and market demand. Indeed, the products mix from a refinery will be different in various seasons because of changing demand. These variations must be assessed on the basis of historical data and cross-country observations, and the forecast must take into account foreseeable changes and process obsolescence.
- Many proposed refinery investments are aimed at rehabilitating and upgrading existing refineries, rather than building new facilities. The economic and financial analysis should be based on a comparison of costs and outputs with and without the proposed investment.

Power Generation

Economic and financial analysis of the power sector usually includes a least-cost analysis of power generating capacity (fig. 9–7), as well as a calculation of economic and financial ratios. The objective of least-cost analysis is to arrive at a composition of generating capacity that would minimize the overall cost of power supply while satisfying demand within technical and environmental constraints. The core computations underlying least-cost analysis are performed by computer packages that are often based on simulation and optimization algorithms. In these computations, the algorithm takes into account growth in demand, shape of the load curve (variations of demand over 24 hours, days of the week, seasons, etc.), costs of various fuels, technical constraints in loading various plants, and so on. The model then calculates the amount and type of generating capacity that should be added to the system over the planning period.

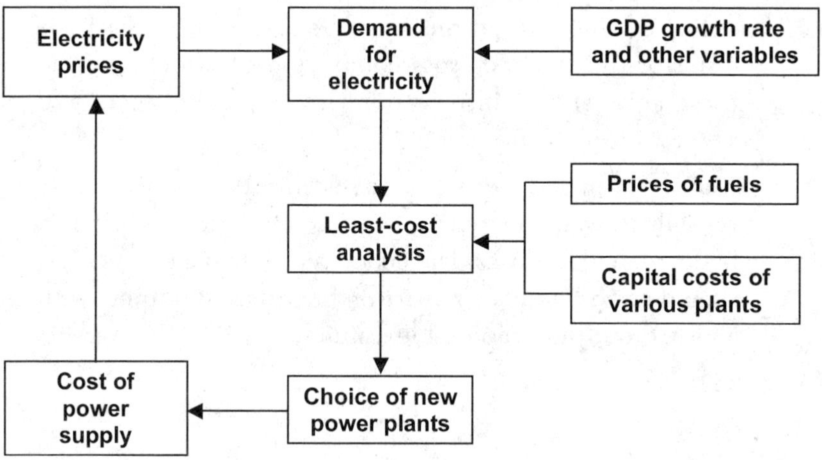

Fig. 9–7. Power generation: The framework of least-cost analysis

Although least-cost computational packages facilitate the analysis substantially, it is important to ensure a broad perspective by considering significant links and by using strategic judgment when interpreting the results. To account for significant links, it is necessary to go through several iterations to

- Assess the cost of power supply
- Estimate the future electricity price
- Examine the impact of the price on electricity demand
- Feed back the revised demand forecast into the least-cost analysis

The Relevance of Average Levelized Cost

In the economic and financial analysis of power generation, it is helpful to calculate the *average levelized cost* of various plant options (tables 9–5 and 9–6). This measure is also referred to as the *average discounted cost* or, under certain conditions, *average incremental cost*. It provides a simple measure of the average capital and fuel costs for each type of generating plant. The calculation of average levelized cost is performed as follows:

$$\text{Average levelized cost} = \frac{\text{PV of cost stream}}{\text{PV of output stream}}$$

where PV is the present value.

TABLE 9–5. CALCULATION OF AVERAGE LEVELIZED COST

Type of plant = Coal-based steam plant

Capacity (MW)	600	Efficiency	38%
Cost (US$/kW)	1,250	Heat rate (BTU/kWh)	9,000
Plant cost (US$ million)	750	Coal price (US$/ton)	55
Plant life (years)	30	Coal heat content (MMBTU/ton)	26
Plant load factor	75%	O&M cost = $25/kW per year + 0.4¢/kWh	

Year	Capital Cost ($ million)	Fuel Cost ($ million)	Other O&M Costs ($ million)	Total Cost ($ million)	Power Generation (GWh)
1	75.00	0.00	0.00	75.00	0.00
2	150.00	0.00	0.00	150.00	0.00
3	225.00	0.00	0.00	225.00	0.00
4	225.00	0.00	0.00	225.00	0.00
5	75.00	75.21	15.77	165.98	3942.00
6	0.00	75.21	15.77	90.98	3942.00
7	0.00	75.21	15.77	90.98	3942.00
8	0.00	75.21	15.77	90.98	3942.00
9	0.00	75.21	15.77	90.98	3942.00
10	0.00	75.21	15.77	90.98	3942.00
11	0.00	75.21	15.77	90.98	3942.00
12	0.00	75.21	15.77	90.98	3942.00
13	0.00	75.21	15.77	90.98	3942.00
14	0.00	75.21	15.77	90.98	3942.00
15	0.00	75.21	15.77	90.98	3942.00
16	0.00	75.21	15.77	90.98	3942.00
17	0.00	75.21	15.77	90.98	3942.00
18	0.00	75.21	15.77	90.98	3942.00
19	0.00	75.21	15.77	90.98	3942.00
20	0.00	75.21	15.77	90.98	3942.00
21	0.00	75.21	15.77	90.98	3942.00
22	0.00	75.21	15.77	90.98	3942.00
23	0.00	75.21	15.77	90.98	3942.00
24	0.00	75.21	15.77	90.98	3942.00
25	0.00	75.21	15.77	90.98	3942.00
26	0.00	75.21	15.77	90.98	3942.00
27	0.00	75.21	15.77	90.98	3942.00
28	0.00	75.21	15.77	90.98	3942.00
29	0.00	75.21	15.77	90.98	3942.00
30	0.00	75.21	15.77	90.98	3942.00
31	0.00	75.21	15.77	90.98	3942.00
32	0.00	75.21	15.77	90.98	3942.00
33	0.00	75.21	15.77	90.98	3942.00
PV @ 10%	561.44	481.33	100.91	1143.68	25227.09

Average capital cost	0.022 $/kWh	=	2.2 ¢/kWh
Average fuel cost	0.019 $/kWh	=	1.9 ¢/kWh
Average O&M cost	0.004 $/kWh	=	0.4 ¢/kWh
Average cost of power generation	0.045 $/kWh	=	4.5 ¢/kWh

TABLE 9–6. COMPARISON OF AVERAGE LEVELIZED COSTS FOR DIFFERENT TYPES OF POWER PLANTS

All in ¢/kWh	Gas-based combined-cycle	Oil-based steam plant	Coal-based steam plant	Integrated gasification combined-cycle	Nuclear plant
Capital cost	1.1	1.4	2.2	2.7	6.6
Fuel cost	2.9	6.0	1.9	1.8	0.5
O&M cost	0.2	0.2	0.4	0.3	0.1
Average cost of power generation	4.2	7.6	4.5	4.7	7.2
Assumptions					
Capacity (MW)	300	300	600	550	1000
Cost ($/kW)	600	800	1250	1500	3000
Plant cost ($ million)	180	240	750	825	3000
Plant life (years)	30	30	30	30	30
Efficiency (%)	49%	38%	38%	41%	33%
Heat rate (BTU/kWh)	7000	9000	9000	8300	10,400
Fuel price	$4.2/MCF	$40/bl	$55/ton	$55/ton	0.5¢/kWh
Heat content	1 MMBTU/MCF	6 MMBTU/bl	26 MMBTU/ton	26 MMBTU/ton	
O&M cost	$12/kW per year + 0.2¢/kWh	$15/kW per year + 0.2¢/kWh	$25/kW per year + 0.4¢/kWh	$35/kW per year +0.3¢/kWh	$16/kW per year + 0.1¢/kWh

All plants are based on 4 years of implementation time, except the nuclear plant which is based on 8 years.
All plants are assumed to have load factors of 75%.
Price for heavy fuel oil based on 80% of a crude oil price of $60/bbl.

Calculation of IRR

Calculation of financial IRR. *Financial IRR* is based on costs and revenues that the project will actually face. Estimates of capital and O&M costs are based on international norms and local conditions. The fuel cost is more difficult to forecast because of the lack of generally accepted forecasts of international prices and because of possible distortions in fuel prices in local markets. The risk associated with fuel costs will be minimized if project sponsors can secure a fuel supply contract with a clear price formula. In some BOT projects, sponsors have managed to shift the risk of fuel cost to the national power company through an arrangement in which the power company provides fuel to the BOT sponsors and receives generated power in return. Under these arrangements, a BOT project functions as a processing unit—receiving fuel and giving back the generated electricity.

The second difficulty encountered in the calculation of financial IRR is forecasting the electricity price. If the project is being built by a national power company, the electricity price estimate should be based on power tariffs and sales to various customer groups. In the case of BOT suppliers, the price risk is minimized through a take-or-pay contract in which project sponsors receive assurances from the power grid system (normally a public utility) that the generated power will be bought based on a clear price formula.

Calculation of economic IRR. Economic IRR is based on costs and benefits of the project to the economy, rather than to the company. The cost figures should be adjusted to exclude taxes and duties and to compensate for other distortions—in exchange rates, fuel prices, and wages.

TABLE 9–7. POWER GENERATION: CALCULATION OF ECONOMIC IRR

	Investment ($ m)				Operation ($ m)								
Year	Gen.	Trans.	Dist.	Total	Gen.	Trans.	Dist.	Fuel	Total	Total Cost ($ m)	Incremtl. Sales (GWh)	Rev. ($ m)	Net Benefit ($ m)
1	36	17.5	20	73.5	0.00	0.0	0.2	0.0	0.20	73.70	0.0	0.00	-73.70
2	54	52.6	20	126.6	0.00	0.0	0.5	0.0	0.50	127.10	0.0	0.00	-127.10
3	54	35.1	25	114.1	3.94	0.0	0.8	2.2	6.94	121.04	22.7	11.52	-109.52
4	36	17.0	26	79.0	3.94	1.6	1.1	26.8	33.44	112.44	273.0	137.52	25.08
5	0	0.0	36	36.0	3.94	1.6	1.5	31.4	38.44	74.44	357.8	180.36	105.92
6	0	0.0	38	38.0	3.94	1.6	2.0	39.4	46.94	84.94	357.8	180.36	95.42
7	0	0.0	35	35.0	3.94	1.6	2.4	40.6	48.54	83.54	535.9	270.18	186.64
8	0	0.0	0	0.0	3.94	1.6	2.4	43.0	50.94	50.94	535.9	270.18	219.24
9	0	0.0	0	0.0	3.94	1.6	2.4	45.2	53.14	53.14	535.9	270.18	217.04
10	0	0.0	0	0.0	3.94	1.6	2.4	47.4	55.34	55.34	535.9	270.18	214.84
11	0	0.0	0	0.0	3.94	1.6	2.4	49.6	57.54	57.54	535.9	270.18	212.64
12	0	0.0	0	0.0	3.94	1.6	2.4	52.0	59.94	59.94	535.9	270.18	210.24
13	0	0.0	0	0.0	3.94	1.6	2.4	55.4	63.34	63.34	535.9	270.18	206.84
14	0	0.0	0	0.0	3.94	1.6	2.4	57.6	65.54	65.54	535.9	270.18	204.64
15	0	0.0	0	0.0	3.94	1.6	2.4	61.0	68.94	68.94	535.9	270.18	201.24
16	0	0.0	0	0.0	3.94	1.6	2.4	63.2	71.14	71.14	535.9	270.18	199.04
17	0	0.0	0	0.0	3.94	1.6	2.4	66.6	74.54	74.54	535.9	270.18	195.64
18	0	0.0	0	0.0	3.94	1.6	2.4	70.0	77.94	77.94	535.9	270.18	192.24
19	0	0.0	0	0.0	3.94	1.6	2.4	73.4	81.34	81.34	535.9	270.18	188.84
20	0	0.0	0	0.0	3.94	1.6	2.4	76.8	84.74	84.74	535.9	270.18	185.44
21	0	0.0	0	0.0	3.94	1.6	2.4	81.2	89.14	89.14	535.9	270.18	181.04
22	0	0.0	0	0.0	3.94	1.6	2.4	81.8	89.74	89.74	535.9	270.18	180.44
23	0	0.0	0	0.0	3.94	1.6	2.4	89.2	97.14	97.14	535.9	270.18	173.04
24	0	0.0	0	0.0	3.94	1.6	2.4	94.0	101.94	101.94	535.9	270.18	168.24
25	0	0.0	0	0.0	3.94	1.6	2.4	94.0	101.94	101.94	535.9	270.18	168.24
26	0	0.0	0	0.0	3.94	1.6	2.4	94.0	101.94	101.94	535.9	270.18	168.24
27	0	0.0	0	0.0	3.94	1.6	2.4	94.0	101.94	101.94	535.9	270.18	168.24
28	0	0.0	0	0.0	3.94	1.6	2.4	94.0	101.94	101.94	535.9	270.18	168.24
PV @ 10%	143	97.4	133	372.9	29.83	10.9	15.9	353.5	410.15	783.05	3281.1	1654.12	871.08
					IRR =	**32%**							

Assumptions:
Generation investment costs based on a 300-MW gas-based combined-cycle plant with a capital cost of $600/kW.
Generation operating costs based on gas-based combined-cycle plant costs of $12/kW per year + 0.2¢/kWh.
Gen. = generation; Trans. = transmission; Dist. = distribution; Incremtl. = incremental; Rev. = revenue.

The benefit stream should be based on the "true" economic benefit of electricity to consumers; however, two difficulties confound assessment of true benefit. First, power tariffs normally underestimate these benefits. An attempt is sometimes made to assess the true benefit through surveys of various customer groups; these surveys examine the behavior of the customers and estimate the cost imposed on each customer group by interruption or shortage of power. Second, power tariffs are economically meaningful only at the point of consumption. Thus, economic IRR is normally calculated for the entire incremental investment—that is, investment in power generation plus the associated investments required in order to expand the T&D system to carry the additional electricity to the consumer (table 9–7).

Power T&D

Economic and financial analysis of T&D projects is normally performed on the "with and without" basis. In the event that the transmission line were specifically used to carry power from one point to another, the "without" case would involve alternative forms of energy transport. For example, in a country where gas reserves are far from the main centers of power consumption, one option would be to build a power plant near the gas field and a transmission line to move the generated power to the consumption center. The alternative would be to construct the power plant near the consumption center and a pipeline to move the gas from the field to the power plant. Thus, the alternative to the power transmission line is a gas pipeline.

Economic evaluation of the transmission project is therefore undertaken as follows:

- The capital and O&M costs of the transmission line are estimated on the basis of international norms and local conditions.
- The benefit stream would be based on the cost of the alternative—the capital and O&M costs of a gas pipeline—as well as other differences. For example, the power plant may cost more to build if it is near a city. It may also include the value of serving new customers along the transmission line (electric vs. gas).

More often, T&D investments are aimed at expanding or reinforcing existing networks. These investments result in reduction of technical losses, improved system reliability, and increased capacity. The comparison of "with" and "without" cases requires quantification of these benefits. Technical losses for the grid (with and without proposed investments) can be assessed with network simulation packages. The reduction in losses is then valued on the basis of the average power tariff. The value of the improved system reliability is difficult to assess; it is normally estimated on the basis of the change in the probability of power outage and the cost of outage to consumers. The economic value of increased capacity is assessed in relation to the power tariffs (capacity and energy charge) minus the cost of power generation.

Economic and financial analysis of power T&D projects may raise the following additional issues:

- As a special case, the transmission line may be entirely associated with a power plant—that is, it is built specifically to transmit power from a plant to the grid. Economic and financial analysis is then performed for the entire generation and transmission scheme.

- Rural electrification is a special T&D project that transports power from the grid to a specific consumption center. Such a project can be evaluated as an independent investment. The project benefits will be assessed on the basis of the cost of energy to the rural community in the absence of access to electricity.

10 Structuring The Financing Package

Although state-owned and privately owned energy companies may be distinguished sharply in theoretical terms, they converge surprisingly in the way in which they seek to finance energy projects in developing countries. State-owned oil, gas, and power companies, which traditionally financed their projects through government budgets or official government-sponsored borrowing, have now turned to commercial sources of finance, such as commercial banks, private bond placements, and even stock markets. By contrast, private investors, who are supposed to bring in commercial funds, often seek ways to use official sources of finance with some type of government sponsorship; this occurs most readily when private sponsors form joint ventures with state entities.

In a similar way, the distinction between corporate (balance sheet) financing and project financing is becoming less pronounced in developing countries. On the one hand, pure corporate financing is becoming rare. Most corporations do not want to keep project debts on their balance sheets for the entire periods of repayment; they try to move the debt off their balance sheets as soon as possible, particularly after the project construction phases. On the other hand, pure project-based financing has turned out to be very difficult for large energy projects in developing countries. Sponsors can mobilize equity and debt financing more easily when they draw on the creditworthiness of parent companies. Therefore, corporate financing and project-based financing are converging toward an arrangement of limited corporate responsibilities, in the form of limited-recourse financing and other types of sponsorship.

As a result of these developments, project preparation and the associated analysis are becoming similar, regardless whether the project is under state or private ownership and whether the project is supposed to be financed on a corporate or a project basis. However, the extent of the required risk mitigation measures varies with the creditworthiness of involved parties and the role of the state in ownership, construction, and operation of the plant.

Determining the ownership structure is extensively intertwined with the design of the financing structure and security package (fig. 10–1). These items will be continuously modified by an iterative process right until the end of the preparation process and, sometimes, throughout the implementation phase.

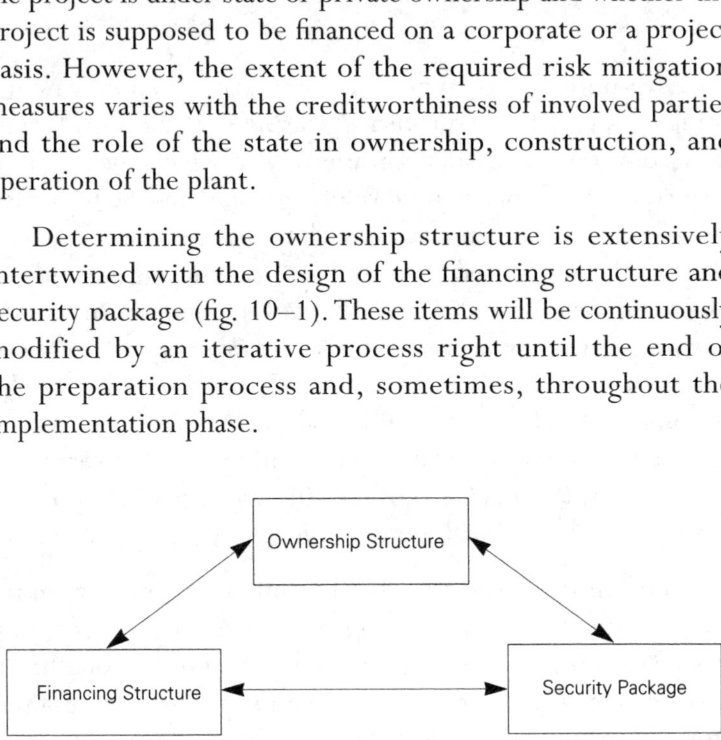

Fig. 10–1. Ownership and financing structures

Determining the Ownership Structure

The ownership structure is determined in two steps: conceptual and technical. The conceptual stage is the time for decisions about the role of various participants—in particular, state versus private and foreign versus local relationships. At the technical stage, the legal structure is devised on the basis of a combination of financial, accounting, and tax considerations.

Types of ownership structure

Ownership structures fall into three broad categories:

- Complete state ownership
- Complete private ownership
- Joint ventures of state and private parties

Complete state ownership. From the point of view of financing, this should be the most convenient ownership structure, because a project undertaken by a state entity can be funded from both public and private sources of finance. Nevertheless, complete state ownership has certain drawbacks.

First, concerns may arise about the efficiency of most state-owned energy entities of developing countries. Such concerns are translated into perceptions of risk regarding construction costs and schedule and regarding plant operation and maintenance.

Second, most multilateral and bilateral financiers discourage full state ownership of energy projects and emphasize private sector participation. As a result, many financiers hesitate to fund projects fully owned by the state; the magnitude and consequence of this hesitation varies with the host-country circumstances and the type of energy project. For most developing countries, the state has ownership of oil and gas resources and often of the upstream oil and gas projects. Although public ownership of resources is often dictated by the law, project ownership can be assigned to the private sector. State ownership is not expected in downstream activities such as refining and product distribution. By contrast, full state ownership is justified in the case of gas infrastructure and power transmission and distribution (T&D). In regard to power generation, private power producers are accepted worldwide as an alternative supply to electricity generated by state utilities; nevertheless, there is still substantial room for generation by state utilities, particularly when it comes to projects with long implementation periods, such as hydroelectric plants.

Complete private ownership. Under this structure, a project is owned by a special-purpose company (a project company) or credible corporations. The international community has demonstrated support for this type of ownership structure in most segments of the energy sector—in particular, upstream and downstream petroleum investments. Nevertheless, several drawbacks are apparent.

For gas infrastructure and power T&D, the natural monopoly characteristics of these facilities must be addressed through regulations. Even in the case of power generation, where private ownership is common and accepted, numerous regulatory matters must be resolved. The challenges of undertaking a wholly private project are more manageable when project sponsors are reputable companies willing to accept formal obligations for efficient construction and operation of project facilities. In addition, direct guarantees by such companies facilitate debt financing.

Overall success also depends on the composition of partnership. A composition likely to be successful would involve powerful local partners and international firms with complementary lines of business—for example, an engineering firm, an equipment supplier, and a construction company.

Project finance is used in cases where financing is not dependent on the credit support of the project sponsors. In other words, in project finance, the lender looks primarily to the revenue stream created by a stand-alone project for repayment and to the assets of that project as collateral. That is, the borrowing is on a nonrecourse basis.

This arrangement is in contrast with corporate financing, where there is recourse to the revenues and assets of the project sponsors or parent companies. In the energy sector, corporate finance was the normal practice until the 1980s, when lenders (in particular, commercial banks) took comfort in knowing that repayment of loans came from the parent entity and was backed by the entity's entire balance sheet. The practice has changed in the past 20 years, partly because sponsor entities do not wish to carry the obligations of the project company on their own balance sheet and partly because there are now numerous players involved in an energy project and each sponsor wants to avoid broader risks, which could be out of its own control.

The first step in project financing is to set up a *special-purpose vehicle* (SPV), as a separate company from the sponsors' entities. The sponsors of the project company usually provide the seed equity capital for the SPV. Another distinguishing feature of project financing is its requirement of a variety of contractual arrangements. Because everything rides on the success of the project, there is a need to deal explicitly with all possibilities of failure—that is, various types of risks. There is a need for numerous contractual arrangements that address every major identified risk associated with the project and allocate each risk to the party that is best able to appraise and control that risk.

Joint ventures of state and private parties. This type of ownership structure is becoming increasingly popular because it provides a very flexible framework for sharing project risks and a wide range of options for equity and debt financing. The most advantageous venture would be between a credible large corporation and an efficient state entity.

It is neither feasible nor desirable to fix the ownership structure at the early stages of project preparation. The ownership structure is determined through an iterative process in which the initial sponsors assess the costs and benefits of involving various sources of equity, finance, and risk management. The optimal structure of an individual project depends on the type of project, the country, and the sponsors' credibility and objectives. However, two common objectives are normally pursued in devising ownership and financing structures:

- To utilize state direct and indirect support as much as possible while minimizing state intervention in the managerial aspects of project construction and operation

- To allocate risks, contributions, and benefits fairly among the interested parties, with due consideration for their respective abilities

Summary 10–1. Determining the Ownership Structure

The ownership structure of a project is determined in two steps: conceptual and technical. The following considerations are important at the conceptual stage:

- Complete state ownership formally provides access to a variety of sources of funds, but in practice, most official sources of funds hesitate to support full state ownership, particularly for upstream oil and gas and downstream oil projects.
- Complete private ownership is supported by almost all players in the energy sector but cannot be easily formulated in the case of gas infrastructure and power transmission and distribution (T&D).
- Joint ventures of public and private sector entities are becoming very popular because of the flexibility they provide for sharing risks and mobilizing debt and equity finance.

At the technical stage, the ownership structure is determined on the basis of the sponsors' tax, accounting, and risk-sharing objectives. Just like conceptual design, technical design of ownership structure is arranged in conjunction with the financing and security structures and addresses the type of company (joint venture, partnership, or single-purpose stock company), place of incorporation or establishment, and so on. These arrangements should be devised to

- Optimize the tax position of the sponsors, taking into account both host- and home-country tax rules
- Achieve the desired risk-sharing objectives of the sponsors
- Maximize the level of nonrecourse financing at lowest cost
- Provide protection for individual sponsors against default by other sponsors

The ownership structure should also include a choice between whether each partner is to be individually responsible for financing, as is typical in oil projects, or whether finance is to be raised for the venture as a whole (i.e., common financing), backed by several undertakings of the sponsors and appropriate interpartner default clauses. Common financing is normally used in the power sector.

Identifying Sources of Finance

Part II of this book describes the background, methods, and conditions of accessing equity and debt finance from multilateral, bilateral, and commercial sources. For most energy projects, sponsors start their investigation of equity and debt financing by considering the entire menu of options. Nevertheless, identification of sources appropriate to the project requires a systematic procedure, as explained in the following subsections.

Assume a flexible ownership structure

A flexible ownership structure is normally assumed by envisaging participation by both foreign and local and both private and state partners.

Examine the government interest

The government interest in the project should be examined at several junctures, with the objective of answering two distinct questions: First, would the government provide or sponsor equity or debt financing under any circumstances? Second, would the government provide administrative support and facility in case it does not wish to be involved in financing? Even a simple declaration of national interest from a "responsible" government agency may help, especially in bilateral negotiations.

Assess the receptivity of the multilaterals

Multilateral aid agencies may not be interested in certain projects or countries regardless of the sponsorship for the project. If the multilaterals do indicate interest, project sponsors should consider an ownership structure that accommodates their requirements.

Approach the bilaterals

Sponsors should approach bilateral aid agencies of their own countries, with the objectives of getting support for project preparation and securing finance in the form of loans or loan guarantees. In addition, they should approach bilateral agencies of other countries that contain well-known suppliers of goods and services needed for the project. Some bilaterals provide investment financing in return for a commitment that the output of certain projects will be imported into their countries. For example, if the output of an oil or gas project can be allocated for export to Japan, project sponsors may be able to obtain loan or loan guarantees from JBIC.

Ask equipment suppliers to mobilize finance

As soon as the overall scope and design of the project are known, project sponsors can identify the most important sources of equipment supply and construction and engineering services. These businesses are often reserved for project sponsors or their affiliated companies. Still, significant components of the project may be acquired from international manufacturers and contractors. Large equipment supply companies, including trading houses, may contribute funds from their own resources. More importantly, they have substantial influence with export credit agencies of their own countries and can mobilize loans or guarantees from them.

Use the IFC or regional development banks to mobilize commercial funds

Large corporations, such as major oil companies, are normally able to access commercial sources of equity and debt finance directly. Other, weaker project sponsors have difficulty doing so on their own. Institutions such as the IFC and the regional multilaterals have experience in quickly assessing the applicability and viability of various arrangements. They can assist by providing and mobilizing equity from institutional investors. They can also structure syndicated loans and mobilize borrowing on bond markets. However, sponsors should be aware that the IFC, like other financial advisors, charges for the cost of its advisory services.

Review the availability of specialized energy funds

These funds, either government sponsored or private, may or may not be available for the project and the country of concern. Nevertheless, if available, these funds can facilitate access to both equity and debt financing.

Survey domestic investors and lenders

Domestic investment and debt funds have become an important source of energy finance, at least in some developing countries. In a number of countries, domestic bond markets have grown rapidly in recent years and have reached, in certain cases, the size and liquidity that can support large-size issues suitable for energy projects. Raising funds on domestic bond markets is easier when there is government involvement or support for the project.

Domestic commercial banks have in recent years shown interest in lending to energy projects. Often, these banks feel more comfortable participating in syndicated loans that would limit the size of their involvement and their relative exposure. The domestic equity markets can also be tapped, under certain circumstances, to fund energy projects.

Finally, in some countries, one can access resources from private companies, private funds, and rich individuals who are willing to take calculated risks by participating in well-designed ventures. Their participation can also facilitate preparatory work of the project because of their familiarity with and influence in the bureaucracy and other local affairs.

Preparing the Security Package

The security package comprises key agreements, contracts, and government undertakings that reduce lenders' and investors' risks by establishing legally binding obligations and procedures. The package includes documents that represent the standard paperwork prepared for a project regardless of who the lenders and investors are—for example, land use and operating license, construction permit, import license, corporate documents, trust agreements, concession agreement, production-sharing agreement, and off-take agreement.

In addition, various guarantees need to be designed into the security package to mitigate project risks. These depend on whether the major portion of debt is financed by commercial or noncommercial lenders.

From a lender's point of view, three questions need fully satisfactory answers:

- Can the project be constructed and commissioned within the planned schedule and budget?
- Can the project generate the projected net revenue?
- Can the net revenue be allocated and paid back to the lenders and investors according to the project agreement?

In association with these questions, lenders want to know who would be responsible for damages in the event of project failure in any of these areas. The security package should provide the lenders with answers to these questions, as well as the means to perfect and enforce their security interest.

Preparation of the security package significantly affects—and is affected by—the ownership and financing structures. In particular, participation by the government in ownership or participation by multilaterals in financing would reduce concerns about political risk and sometimes increase concerns about commercial risks. The point is that the security package will continue to be modified until the ownership and financing structures are finalized. Preparing an appropriate security package requires a systematic procedure, as explained in the following subsections.

Identify sources of applicable guarantees

It is not feasible or advisable to ask for commitment from any commercial or official sources of guarantee before deciding about partners and financiers. A guarantor would need to know other important players. In addition, the structure of financing affects the options for procurement of goods and services, which in turn affects the choice of guarantor. Nevertheless, it is possible to compile a list of potential guarantors and inquire about their interest in participating in the proposed project. This information is needed before approaching financiers.

Evaluate options for mitigation of political risks

Although mitigation of commercial risks represents the most time-consuming aspect of preparing the security package, the issue of political risk should be addressed at the outset. Most investors and financiers are convinced that commercial risks can be effectively addressed when the time comes, but they also feel that political risks are out of everybody's control. Thus, they do not take a proposal seriously until they receive some comfort that political risks are manageable. Options for mitigating political risks include several types of guarantees, as well as the involvement of certain players—for example, partnership with a key state entity or with powerful local individuals and companies.

Formal guarantees against political risk are provided by the host government and by multilateral and bilateral agencies. Often, the question is not about choosing one versus another but about combining them to provide the most comprehensive coverage at the lowest possible cost. Although it is a partial guarantee in name, the World Bank's guarantee instrument is relatively broad and should be viewed as a good starting point in sponsors' investigations for political risk guarantee. The World Bank also provides leads to other applicable sources of guarantee. Other multilateral institutions—for example, the regional banks—also provide guarantees and assistance in identifying sources of guarantee. Most multilaterals have regular working relationships with the bilateral agencies, know of their country and sectoral preferences, and cooperate with them in processing requests for loans and guarantees.

Investigate options for mitigation of commercial risks

This is certainly the most time-consuming step in the preparation of the security package. The task has two distinct dimensions. First, sponsors need to reach agreement with the government or certain government entities about aspects of marketing the project's output. The government's role depends on the country and the type of project. However, for gas and power projects, the government's role is substantial, because the output is primarily bought by a state entity or because prices are regulated by the state. Therefore, project sponsors need to secure take-or-pay or throughput agreements with the state entities. The government would need to guarantee either the credibility of the state entities or that it will permit a sufficient increase in energy prices.

Securing government guarantees and agreements takes a relatively long time to complete, particularly in countries that lack clear precedents.

The second dimension in mitigating commercial risk involves negotiating with contractors, equipment suppliers, fuel suppliers, operating companies, and so on, to determine their willingness to compensate for damages if they fail to fulfill their obligations. This process is technically complex but is normally accomplished efficiently because it is driven by commercial incentives. The prerequisite for completion of these negotiations is a firm plan for procurement of goods and services.

Finally, the security package should provide details regarding what happens in the event of a default or bankruptcy of the project company. The security package normally includes pledges of the project's assets to the lenders in accordance with a specific order of debt seniority, in the event of default. However, many local legal and regulatory issues have to be incorporated into the pledge. In some countries, private and/or foreign acquisition of natural resources or even real estate may not be legally permitted. Another issue of concern in energy projects is that pledges of assets such as a power plant or a gas pipeline may not have real value to the lenders.

Summary 10–2. Preparing the Security Package

- The security package includes some standard documents (e.g., permits and licenses) and various guarantees or agreements aimed at (1) managing and mitigating risk, (2) enabling the pledge of sponsors' assets, and (3) perfecting and enforcing the security interest.
- Contents of the security package will be modified continuously through the preparation process because they are affected by the ownership and financing structures. In particular, an increased government role in the project may reduce the need for arranging insurance and guarantees against political risk but may also increase the need for guarantees against commercial risks.
- Although mitigation of all risks will have to be considered, the issue of political risk should be addressed at the outset. Most investors and financiers do not take a proposal seriously until they receive some comfort that political risks are manageable. Among various commercial risks, handling the market risk and securing the corresponding guarantee instrument, such as a take-or-pay contract, are the most challenging. Somewhere between political and commercial risks is the currency exchange risk.

BOT Schemes

Describing a BOT scheme is a useful exercise that brings together the various topics of this chapter. The underlying arrangement for a BOT scheme is a concession given to a company to provide an infrastructure service for a certain period of time. While such arrangements have been in practice for a long time, the term *BOT* has been used in the industry only since the 1980s. Early in its usage, emphasis was placed on the *transfer* part of the term, because of political sensitivity attached to and, sometimes, legal barriers opposing private ownership of infrastructure facilities—in particular, energy assets. A BOT scheme therefore assured the public that the ownership of a facility would eventually revert to the state.

Over the past two decades, the BOT concept has become more flexible, allowing the private sector to retain ownership of facilities through BOO schemes (and a number of variations with acronyms that reflect their distinctions). Also over the course of this period, the BOT concept has fluctuated in popularity and feasibility across countries and regions.

Today, the terms *BOT* and *BOO* are used to refer to concessions for private sector participation in the energy infrastructure services. Although BOT schemes are concentrated in power generation, their application can be extended to almost every segment of the energy sector.

The major parties involved in a BOT scheme are as follows:

- The government (or a government agency that awards the concession contract to the BOT company). The process of awarding the concession takes various forms. For most power projects, the government uses competitive bidding to select among the interested companies. The basis of the bidding is often the price of the delivered service. For example, the company that proposes the lowest price for electricity delivered would be considered the winning bidder. However, the basis of bidding could vary depending on the objective of the concession. For example, if the government intends to supply electricity to rural areas at a predetermined price, the bidders could compete on the level of required subsidy from the government.

- Promoter of the BOT, the entity (company) that has been granted the concession to build, operate, and transfer the facility to the government when the concession contract expires. The promoter usually establishes a special-purpose company to undertake the project. The project company will enter into contractual arrangements with relevant parties to finance, build, and operate the plant.
- Investors/shareholders, normally including the promoting sponsors, among others. They provide equity and receive dividends as return on their investment. Some shareholders may have direct interest in the project (e.g., contractors and fuel suppliers) while others may be equity investors (e.g., public shareholders and institutional investors). Shareholders could also include government agencies. An interesting example of government agency participation is the Powerlinks BOT project in India, where the national grid company and a private utility (Tata Power Corporation)

have formed a project company to build and operate a power transmission system (for discussion of this project, see the final section, "Examples of Financing Structures").

- Lenders, including multilateral, bilateral, and commercial entities that provide loans on a nonrecourse or a limited-recourse basis. In most BOT projects, there are a number of lenders, one of which plays the role of lead lender.
- Purchaser of output, often a public utility, such as the national electricity grid company or the national gas company. The purchaser could also be a power or gas distribution company or consumers at large.
- Suppliers of fuel or other input that could be private or state-owned company. A BOT project in power generation often receives its fuel from a state-owned company. A BOT project in power transmission receives its input (electricity) from a public or a private company.
- Construction company (a single company or a consortium of companies). In some projects, the promoter might be or might include a construction company.
- Plant operation company. An additional entity is involved when the facility is not operated by the promoter but has been contracted out to a specialized operation and maintenance company.

A mirror image of the preceding list of parties is presented by the following list of contractual agreements that are required in order to determine the exact terms of relationships among the parties:

- Concession contract, which determines the terms and conditions under which the promoter is granted the right to supply a service. This contract describes the responsibilities for financing, construction, operations, and maintenance, as well as the condition of the facility at the time of transfer to the government when the concession expires.
- Shareholder agreements, which determine the responsibilities and relationships among the shareholders including the obligations in the event of loan default or project failure.
- Loan agreements, between the project company, lenders, loan underwriters, guarantors, and insurers.
- Off-take contract, which obligates the purchaser(s) of the output to buy the entire or a portion of the output at a fixed price (or based on a price formula) during a specified period of time. A power purchase agreement is a famous type of off-take contract.
- Fuel or input supply contract, which obligates a company to provide fuel at a fixed price (or according to a price formula) during a specified period of time.
- Construction contract, which might be a turnkey contract or another type of contract. The construction contact determines the time, budget, and standards of quality within which the project would be implemented. The contract should include provisions for possible failure in any of these areas.
- O&M contract, which determines the budget, performance, and maintenance standards, as well as provisions for any possible failure.

Summary 10–3. Build-Operate-Transfer Schemes

- The term *BOT* (build-operate-transfer) has been used in the energy industry since the 1980s to refer to concessions given to the private sector to provide an infrastructure service.
- The emphasis was initially on the word *transfer,* to cope with the political sensitivity and legal barriers to private ownership of energy sector assets.
- The major parties in a BOT scheme are the promoter, investors, the government, lenders, the purchaser of output, the supplier of input (fuel), the construction company, and the operations and maintenance (O&M) company.
- The preceding list is mirrored by the following list of required contractual arrangements: concession contract, shareholders agreement, loan agreements, off-take contract, construction contract, and O&M contract. Some of the agreements and contracts may be backed by guarantees and insurances from specialized entities (fig. 10–2).

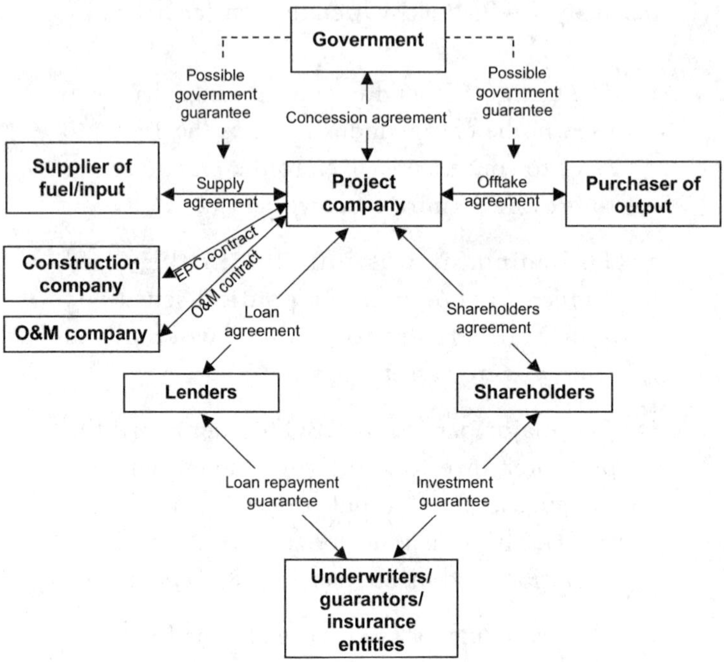

Fig. 10–2. Schematic presentation of contractual relationships in a BOT transaction

Selecting Private Power Producers

In the 1990s, a number of developing countries learned, through a process of trial and error, how to award concessions for infrastructure services. Power was revealed to be a high-profile sector—in terms of volume, as well as commercial interest and political issues that were raised during and after concession award. Countries such as Chile approached private power in a calculated and transparent manner. However, countries like India, Pakistan, Turkey, Thailand, and Indonesia jumped into private power and subsequently faced serious problems that harmed the creditworthiness of the countries and of the private companies. One of the most notable examples of this learning curve was private power in Pakistan, which has become a classic case that other countries study to avoid potential problems in the future.

Pakistan was one of the pioneers of private power and paid a high political and economic price owing to its many mistakes. Its first mistake was to count on the supply of the private power prematurely. Subsequent to a high-profile but poorly designed initiative for private power in the 1980s, the government of Pakistan cancelled plans for construction of power plants by the public sector. Finalizing the transaction for the first private power plant took far longer than initially expected. As a result, the country faced serious power shortages, which imposed substantial cost on the economy. The first private power scheme—the famous Hub Power project (1,292 MW; $1.6 billion)—was politically controversial but, after taking years to materialize, was eventually brought to financial closure. Despite all difficulties, this scheme provided a means to formulate energy policy and work out the details of purchasing private power. At the time of its closure in the mid-1990s, the Hub Power project was named the Deal of the Year by *Euromoney Institutional Investor.* Subsequently, the government of Pakistan signed contracts for 19 other independent power producer (IPP) transactions and was praised by the international financial and energy community as a model for private power. However, by the late 1990s, the government was forced to issue notices of intent to terminate 11 IPP agreements and investigate the others.

Ultimately, the investigation into and the political dispute over private power (including the Hub power project) was intense, complicated, and frustrating for all parties. Criminal allegations (of corruption, bribery, and kickbacks) were mixed up with commercial issues (tariffs, financial damages, etc.) and with technical aspects of the projects in question. It took several years of struggle and negotiations for Pakistan to resolve the IPP disputes. The lessons that emerged from the process are similar to the lessons learned from experiences in Turkey, Indonesia, Thailand, and some countries in Latin America.

The first problem was that too many IPP contracts were signed in a short period of time and were coming onstream, simultaneously, at a time that demand for power did not grow as anticipated. This was exacerbated by the national power company's selling the electricity to the public at a price lower than the price at which it was buying the electricity from the IPPs.

The second problem was that most IPP plants employed inefficient diesel or gas turbines that were being built because they involved smaller up-front investment and shorter implementation time. As a result, the emerging mix of power generation technology was not desirable from either an economic or a technical point of view.

The third problem, which turned the entire matter in a troublesome direction, was the bidding or selection process. The Hub Power project was a negotiated deal that had gone through many rounds of revisions prior to reaching financial closure. Therefore, in retrospect, the public could not see, in a transparent manner, the basis of the final price. Other IPP contracts had been awarded to proposals that met the price ceiling of 6.1¢/kWh. This latter approach did portray some degree of transparency but was viewed at a later stage as being too expensive.

Today, most BOT awards are based on competitive bidding. In this manner, the government determines the overall parameters of the desired facility (including the capacity and the technology) and invites the interested parties to bid based on a clear measure, often the price that they would charge for the service. It is also becoming common practice to limit the take-or-pay contract to a certain percentage of the output and/or to certain parts of the concession period. Also, in a new development, the number of BOT transactions, which had dropped significantly between 1998 and 2002, has been on the rise since 2004. A number of these transactions are presented in the next section, which describes various energy project financing structures.

Examples of Financing Structures

Although it is not feasible to present standard models of energy project financing structures, historical experience is instructive. The examples discussed in this section were chosen to illustrate various types of arrangements. Together, these examples represent the wide array of instruments available; selecting the right combination for the prevailing circumstances is an art. Many of the examples correspond with cases that reached financial closure during the 2003–2006 period. There are also a few older cases that have been included because of their unique features or reputation in the international financial community.

FINANCING UPSTREAM OIL AND GAS PROJECTS

Upstream oil and gas projects in developing countries are normally developed either by the country's NOC or by ventures involving IOCs. Financing for these projects is usually based on internal cash generation and corporate borrowing. This is particularly true of projects developed by large IOCs, which utilize a corporate pool of funds mobilized through a wide range of financial instruments and markets. For many smaller IOCs, corporate borrowing is limited by the strength of their balance sheets. They often borrow funds on nonrecourse or limited-recourse bases. The following examples indicate the types of financing structures for state-owned, private, and public/private joint venture projects.

SICHUAN GAS DEVELOPMENT PROJECT. Review of the Sichuan gas project demonstrates how a state-owned upstream project can be financed. It also indicates how various sources of grants can be utilized to finance project preparation and implementation.

Sichuan is the most populous province in China. It was also the country's major source of gas supply. However, gas production declined rapidly because of aging wells. A major project to reverse the decline was formulated in the mid-1990s and implemented during the period of 1997–2003. The project cost was $945 million and was funded by government support.

Responsibility for gas production, distribution, and transmission lies with the Sichuan Petroleum Administration (SPA), a subsidiary of CNPC. The project financing was to a large extent dependent on SPA's own resources but was also supplemented by CNPC's internal cash and government-guaranteed official borrowing from the World Bank (fig. 10–3).

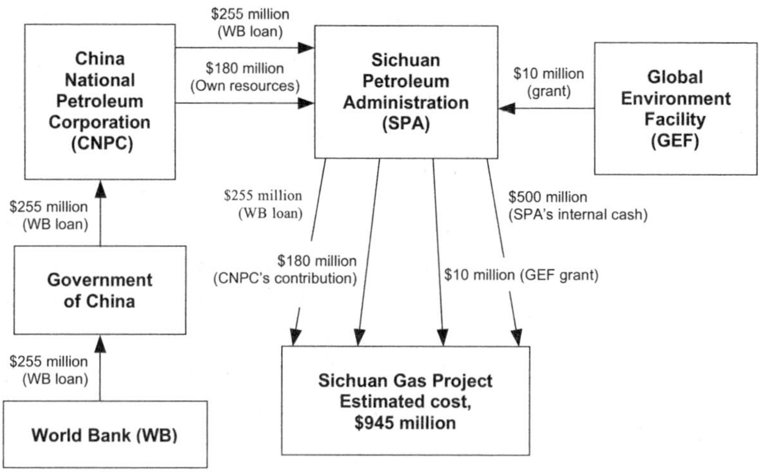

Fig. 10–3. Financing structure of the Sichuan gas development project

The project financing plan was as follows:

SPA internal cash	$500 million
CNPC internal resources	$180 million
GEF	$10 million
World Bank	$255 million
Total	**$945 million**

Although the financing structure is simple, it includes interesting features. First, preparation of the project was supported by grant funds from Japan, the UNDP, GEF, and others. Because of the environmental advantage of natural gas over coal and because Sichuan is a poor province, the international community had substantial sympathy for the project. Second, project implementation was also partly financed by a GEF grant of $10 million. The justification for GEF grant funding was that rehabilitation of SPA's T&D system is expected to reduce leaks of natural gas, whose main component, methane, is a potent greenhouse gas.

Qatar Gas III. Qatar has aggressively developed its gas production capacity. Although the initial phases of Qatar's gas development proved hard to implement owing to challenges of mobilizing financial resources, its more recent projects have moved very smoothly. Qatar Gas III was initiated in 2005, with a target commissioning date of mid-2009. It is an integrated project comprising

- Upstream gas production facilities, to produce approximately 1.4 billion cubic feet per day of natural gas and 70,000 b/d of LPG and condensate from the North Field over 25 years
- Two large-scale LNG trains, each with a capacity of 7.8 million tons per year

The project is jointly owned by Qatar Petroleum (68.5%), ConocoPhillips (30%), and Mitsui (1.5%).

Qatar Gas III is funded by 31% equity and 69% debt (fig. 10–4). The equity amount, totaling $1.8 billion, was contributed by project sponsors in proportion to their shares. The debt amount, totaling $4 billion, was rapidly mobilized in 2006 with favorable terms. JBIC provided a $1.0 billion loan in support of the objective of the project and the participation of Japanese firms. USExim provided $300 million with a similar objective. A significant portion of the output of the project is expected to be shipped to U.S. markets through ConocoPhillips. Another interesting feature of this project's financing is the lending facility of $1.2 billion by a wholly owned subsidiary of ConocoPhillips. Finally, the commercial loan facility of $1,488 million was arranged as a syndicate of 26 international, regional, and Qatari banks. All facilities have a maturity of 16 years, which is a favorable term for this type of borrowing.

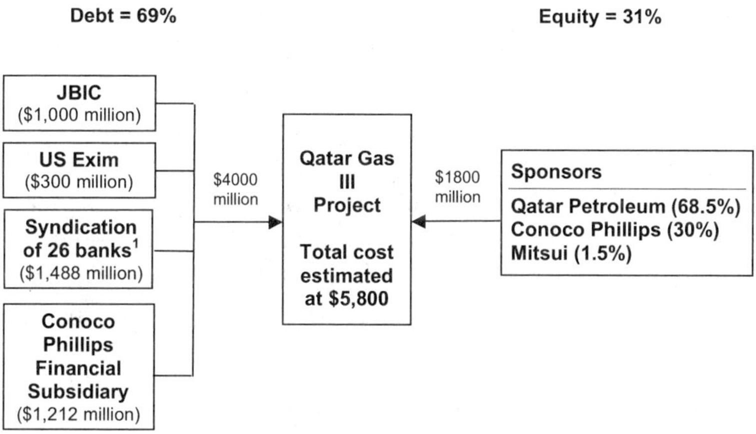

[1]Banks include: Apicorp, BNP Paribas, Mizuho, Qatar National Bank, Arab Banking Corporation, Bank of Ireland, Bank of Tokyo-Mitsubishi, BayernLB, Calyon, Citigroup, Commercial Bank of Qatar, Dexia, DnB NOR Bank, Export Development Canada, Fortis Bank, Gulf International Bank, HSBC, ING, Mashreqbank, Natexis Banques Populaires, Royal Bank of Scotland, Sanpaolo IMI, Standard Chartered, SMBC and WestLB

Fig. 10–4. Financing structure of Qatar Gas III

Tangguh LNG plant. Financing of the Tangguh plant, in Indonesia, provides an interesting example of how various commercial interests have merged to support a very big and complex project in a country where political and commercial risks are high. The project cost, amounting to about $5 billion, is funded primarily through commercial sources, with the support of the government, JBIC, and the ADB. The project has been under discussion since the late 1990s but received financing only in 2006, when many interested parties were seeking additional sources of gas supply and opportunities to lend in a very competitive market. The financing structure has also led to an unexpectedly low-cost debt financing.

The Tangguh project consists of a two-train, 7.6-million-ton-per-year LNG liquefaction project in Papu, Indonesia. The project was initiated in 1997 by ARCO and Petramina, when the Tangguh gas fields were discovered. Presently, BP is the operator of the Tangguh project. The shareholders of the project are BP, the Chinese National Oil and Gas Company (CNOOC), Mitsubishi, Inpex, Nippon Oil Kanematsu and Japan Oil, and Sumitomo and Sojitz. The output of the project is sold through three off-take contracts to China (through CNOOC and the Fijian LNG terminal), Korea (through K-Power and Posco), and the United States (through Sempra Energy), in volumes of 2.6 million tons, 600,000 tons, and 3.7 million tons per year, respectively.

The financial structure of the Tangguh project is based on 30% equity and 70% debt (fig. 10–5). The equity amount, $1.5 billion, is provided by project sponsors in proportion to their shares in the project. The debt amount, $3.5 billion, has been mobilized through four arrangements:

- JBIC provides $1.2 billion in line with Japan's general policy of securing energy supply to the Asia region.
- The ADB provides $350 million from its private sector arm without requiring the guarantee of the government of Indonesia.
- A syndication of seven international commercial banks (Bank of Tokyo-Mitsubishi, BNP Paribas, Fortis Bank, ING, Mizuho, Standard Chartered, and Sumitomo Mitsui Banking Group) lend $1,050 million. The terms of this syndicated loan are surprisingly favorable. For a 15-year

term, the syndicated loan carries an interest rate of LIBOR plus 22 bp, which is much lower than one would expect for a loan of this duration and project location. However, this syndicated loan, as well as the loans by JBIC and the ADB, is guaranteed by BP and CNOOC.

- The fourth arrangement includes a syndication of five Chinese banks (Agricultural Bank of China, China Construction Bank, China Development Bank, Eximbank of China, and Industrial and Commercial Bank of China). This syndicated loan is not guaranteed by project sponsors but is linked to the off-take contract that has been negotiated between the governments of Indonesia and China.

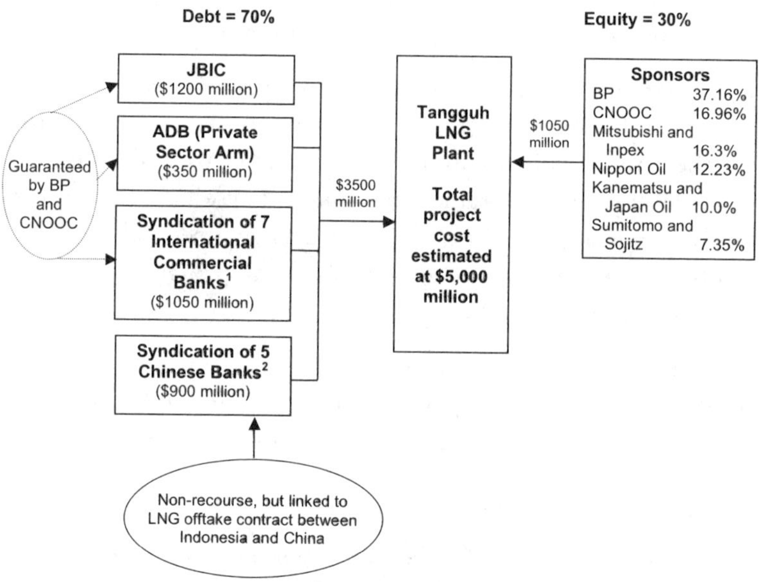

[1] Bank of Tokyo-Mitsubishi, BNP Paribas, Fortis Bank, ING, Mizuho, Standard Chartered, Sumiitomo Mitsui Banking Group
[2] Agriculture Bank of China, China Construction Bank, China Development Bank, Eximbank of China, Industrial and Commercial Bank of China

Fig. 10–5. Financing structure of the Tangguh LNG plant

Guangdong regasification terminal. Financing of the $900 million Guangdong regasification terminal demonstrates the Chinese ability to attract new technology and finance and implement a project in an efficient manner. The project involves the construction of China's first LNG receiving terminal and its associated high-pressure gas pipeline, to supply the Guangdong province with 4 billion cubic meters/year of natural gas (equivalent to 3.3 million tons/year of LNG). The project was launched in 2002 and completed in 2006. The project sponsors were CNOOC (33%), Guangdong Province Consortium (31%), BP Amoco (30%), Hong Kong Electric and Light Company (3%), and Hong Kong and China Gas Corporation (3%). The LNG supply is secured through a contract with the North West Shelf Venture, an Australian-based consortium, for 25 years (3.33 million tons/year).

Financing of the Guangdong project is based on 27% equity and 73% debt (fig. 10–6). The equity is contributed by the sponsors in proportion to their shares. The debt is provided by Chinese banks on a nonrecourse basis. However, the Guangdong project company has secured take-or-pay contracts with major gas consumers in the province to ensure its projected stream of revenue.

Baku-Tbilisi-Ceyhan pipeline. Financing of the Baku-Tbilisi-Ceyhan pipeline was one of the most challenging transactions in the industry for several reasons:

- Involvement of three countries in the project
- Various physical, including environmental, constraints
- Timing of the project financing—in the late 1990s, when international oil prices were still in the vicinity of $20/bl

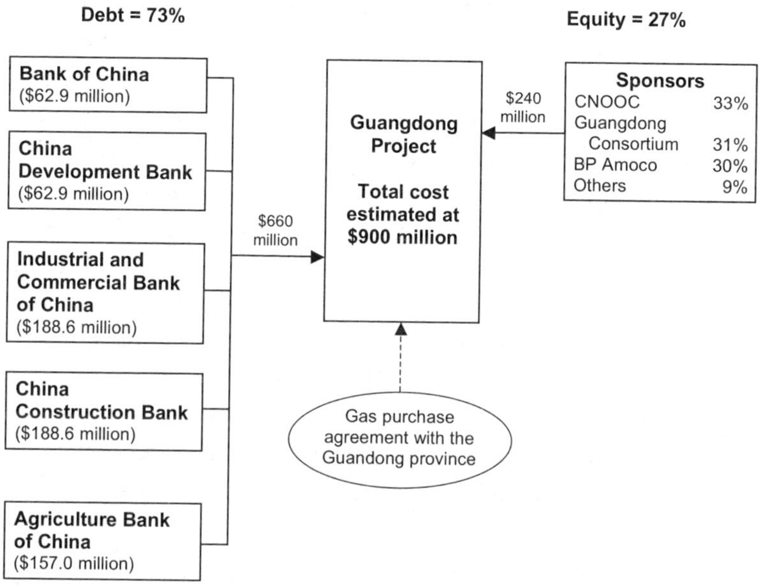

Fig. 10–6. Financing structure of the Guangdong regasification terminal

The Baku-Tbilisi-Ceyhan pipeline was commissioned in 2006 and transports crude oil 1,776 kilometers (km) from the Azeri Chirag-Guneshli field in the Caspian Sea through Georgia and Turkey to Ceyhan, which is a port on the southeastern Mediterranean coast of Turkey. Pipeline construction began in mid-2002 and was completed in 2006. It is expected to reach its full capacity of 1 million b/d by 2008. The project sponsors are BP (30%), SOCAR (25%), UNOCAL (8.9%), Statoil (8.7%), TPAO (6.5%), ENT (5%), Total (5%), and Amerada Hess, ConocoPhillips, Inpex, and Itochu (holding the remaining 10% of shares).

Preparation and attempts to finance the Baku-Tbilisi-Ceyhan pipeline project started in 1999. The project cost was about $3.3 billion. The view at the time was that a project of this size and complexity could not be financed unless fully supported by the corresponding governments. The support was initially meant to include funding. Over the course of almost three years, the project received full funding (fig. 10–7). The equity share of 30% was contributed by the project sponsors. The debt financing was mobilized through very extensive political lobbying by project sponsors and various governments, including the U.S. government. Two multilateral institutions—the IFC and the EBRD—took active roles in defending the project. Seven bilateral entities provided financial support, and a syndication of 11 commercial banks provided loans.

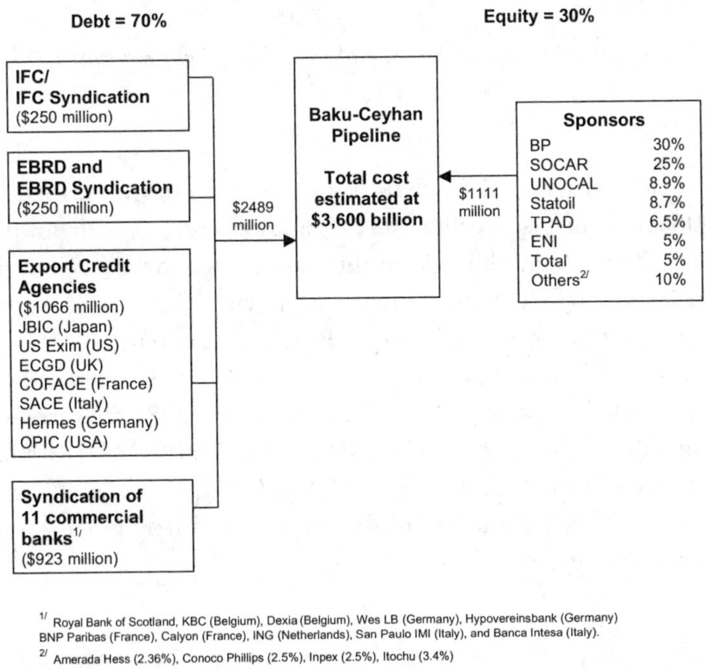

Fig. 10–7. Financing structure of the Baku-Tbilisi-Ceyhan pipeline

Cusiana oil pipeline. The project commonly referred to as Cusiana is based on a development plan for two adjacent crude oil reservoirs, the Cusiana and the Cupiagua fields, in Colombia. There have been various financing schemes associated with various phases of the project. The financing scheme for the pipeline portion has become well known in the industry as an interesting and complex case that was formulated successfully.

The exploration licenses for the Cusiana and Cupiagua fields were initially given to Triton in the early 1980s. Subsequently, BP and Total together formed into these blocks, with BP assuming operatorship. Working interests of the three foreign partners during the exploration and appraisal phases were as follows: BP and Total, 38% each; Triton, 24%. Under a standard provision for contracts in Colombia, Ecopetrol (the Colombian state oil company) exercised its right to take a 50% working interest in the licenses, following a formal declaration of commerciality in 1993. The development project was accordingly undertaken by Ecopetrol (50%), BP and Total (19% each), and Triton (12%). Each partner's production rights are subject to an overriding government royalty of 20%, reducing the net production shares for BP and Total to 15.2% each and for Triton to 9.6%. Each foreign partner is subject to the normal corporation tax regime applicable to industrial activity in Colombia.

Two well-known projects are associated with Cusiana. Phase 1 constitutes development of Cusiana and Cupiagua fields within the limits of existing pipeline capacity. Phase 2 is intended to substantially expand the pipeline capacity and increasing upstream production capacity, to 500 million b/d.

The Phase 1 project, costing about $1 billion, was financed by the partners in proportion to their shares. Each partner took responsibility for mobilizing its own financing. There was no project-based financing except for USExim credits used to fund contributions of two individual partners.

The funding of the pipeline for Phase 2 took a different course and mobilized finance on a project basis. The main reason behind this was a change in government policy in regard to the contribution of public budget to oil projects. In 1994, the government announced its intention that pipeline infrastructures were to be developed by the private sector, rather than under the dominant influence of Ecopetrol. To facilitate this process, the government expressed a preference that the pipeline project be constructed by a joint-stock company in which the foreign partners could, if desired, take shareholdings corresponding to their own transportation requirements.

Accordingly, a capital stock company, Ocensa, was created for the purpose of constructing the Oleoducto Central Pipeline. The new pipeline (about 800 km), was scheduled for implementation in 1996–1997 to transport oil from the Cusiana and Cupiagua fields to the port of Covenas, on the Caribbean coast. Crude oil is then transported from this port to U.S. Gulf and East Coast ports. Ocensa's shareholders are as follows: Ecopetrol, 25%; BP Colombia Pipeline Ltd. (United Kingdom), 15.2%; Total Pipeline Colombie (France), 15.2%; Triton Pipeline Colombia (United States), 9.6%; Interprovincial Pipelines Enterprises (IPL; Canada), 17.5%; and TCPL International (TransCanada), 17.5%. BP, Total, and Triton will use their share of the pipeline system to ship their own oil. IPL and TCPL will ship Ecopetrol's oil.

The financing structure of the Ocensa project is 30% equity and 70% debt financing (fig. 10–8). The equity portion—$608 million—consists of $504 million paid-in cash by shareholders and $104 million expected in cash from tariffs.

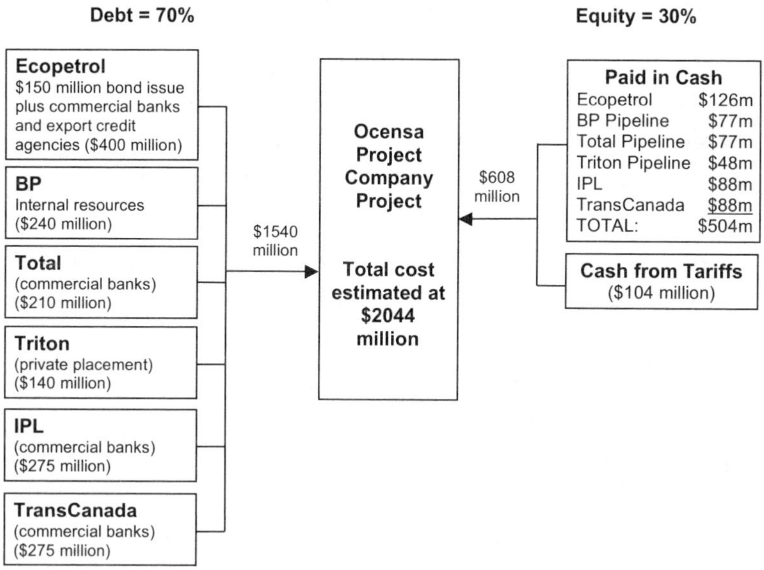

Fig. 10–8. Financing structure of the Ocensa project

The debt is financed through a variety of methods. The actual borrower is Ocensa, but each oil shipper arranged its own portion of debt, according to its interest in the upstream field: Ecopetrol used a combination of bond issues, commercial bank borrowing, and loans from export credit agencies to borrow $400 million; BP continued its practice of funding projects from its corporate resources; Total used a consortium of banks to arrange $210 million in commercial loans; and Triton financed its share of debt through private placement with institutional investors. The borrowing by each company is backed by transportation agreements for shipment of the company's oil through the pipeline system. Ecopetrol's borrowings are backed by a transportation agreement for shipment of Ecopetrol's oil.

Financing petroleum refining projects

Petroleum refineries are built by private companies in most developing countries. Even in countries where a state company owns the refining facilities, new investments are usually funded by commercial sources or a combination of public and commercial funds.

SAMIR refinery upgrade. Financing of the SAMIR (Societe Anonyme Marocaine de l'Industrie du Raffinage) refinery project in Morocco represents an interesting case in which a hybrid of corporate and project finance was used to mobilize about $670 million in debt financing from various domestic and international financiers. SAMIR owns and operates the only two refineries in Morocco.

SAMIR was privatized in 1997. The shares are owned by Corral Petroleum Holdings (67.7%), which is a Saudi firm based in Sweden, and Holmarcom Group (7.1%), which is a Moroccan company, and are publicly traded on the Casablanca Stock Exchange (25.2%). The privatization transaction stipulated the upgrading of the larger refinery (120,000 b/d) at Mohammedia, to meet the EU's environmental standards. The upgrade, which is estimated to cost about $1 billion, will include installation of a vacuum distillation unit, a hydrocracking unit, and a hydrogen and sulfur production unit; these units will allow substantially increased production of gasoline—a 45,000 b/d expansion capacity—while also meeting the EU specification of a low-sulfur product slate.

The financing of the $1 billion upgrade was finalized in 2006, after several years of challenging negotiations (fig. 10–9). About 30% of the funds, or approximately $340 million, are expected to come in the form of equity by project shareholders. Some $82 million of this amount will be in the form of paid-in capital, while the rest will come from the internal cash generation of SAMIR. The debt portion, approximately $670 million, is provided through a combination of four arrangements:

- The AfDB provides $85 million.
- The French development bank, Proparco, provides $30 million.
- The international commercial tranche is arranged by BNP Paribas and Barclays Capital. The total amount is $290 million, of which $240 million is supported by insurance, through SACE, against political and commercial risks. SACE's involvement is in support of the main EPC contractor, a consortium led by the Italian Firm Snamprogetti.
- The domestic commercial tranche amounts to $265 million and is provided by Banque Marocaine du Commerce Exterieur (BMCE), Attijariwafa Bank, and the BCP Bank.

All debt tranches have 13-year tenors, except for the international commercial tranche, which is for 10.5 years.

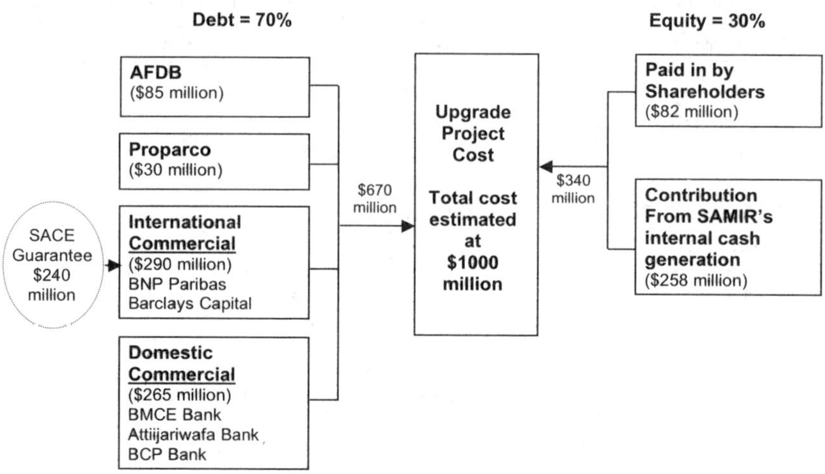

Fig. 10–9. Financing structure of the SAMIR refinery upgrade

Reliance refinery and petrochemical project. Financing of the Reliance refinery and petrochemical project represents an impressive achievement in raising equity and debt funds on a very large scale. The project proposed is the building of a 580,000 b/d refinery and a 900,000 tons/year polypropylene petrochemical complex at Jamnagar, in western India. The project cost is estimated at $6.4 billion; project completion is planned for 2008. The project company is Reliance Petroleum, which is itself a spin-off from Reliance Industries.

Funding for the project was raised in the form of equity and debt (fig. 10–10). Project funding started with a three-part IPO in April 2006. The first part, comprising 450 million shares, was publicly offered. The IPO was oversubscribed significantly and raised $600 million. The second part, another 450 million shares, was sold to institutional investors for another $600 million. The third part, accounting for 900 million shares, was sold to the parent company, Reliance Industries.

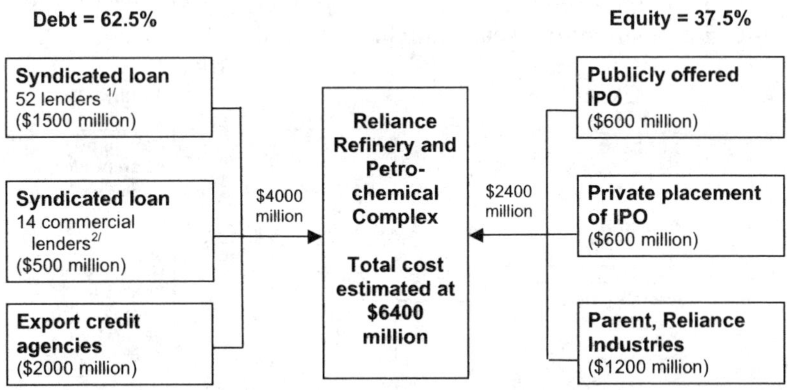

1/ Includes international, regional., and Indian commercial lenders

2/ Includes the MLAs of the transaction: ABN Amro, Bank of America, BNP Paribas, BTMU, Calyon, Citigroup, DBS, DZ Bank, HSBC, ICICI, Mizuho, SMBC, Standard Chartered, State Bank of India

Fig. 10–10. Financing structure of the Reliance refinery and petrochemical project

The debt financing was more challenging because international commercial banks had not provided large nonrecourse loans to India since the $2.5 billion Dabhol power project collapsed in 2001. However, the loan syndication for the Reliance project was received very well by international, regional, and Indian commercial lenders. Overall, 52 commercial lenders participated, providing $1.5 billion. The terms were attractive, including tenors of 7.5 and 10 years and interest margins of 80 bp and 85 bp over LIBOR. The second $500 million loan was provided by a subset of the 52 lenders—namely, the 14 lenders who served as the MLAs. Their contribution was pro rata to their original commitments in the $1.5 billion loan. The remaining $2 billion debt is slated for loans from export credit agencies in connection with equipment and construction contracts.

Yanbu National Petrochemical (YANSAB) project. In 2006, financing of the YANSAB project, in Saudi Arabia, set a new benchmark in volume and speed of equity and debt mobilization. The project is a greenfield development, financing is nonrecourse, and volume is large.

YANSAB is an integrated petrochemical complex to be built in Yanbu, in the western province of Saudi Arabia. It will produce 1.3 million tons per year of ethylene. The total project cost is estimated at $5 billion. SABIC owns 51% of the YANSAB company, while a further 4% is wholly owned by a subsidiary—SABIC Investments Company. Another 10% of the equity is held by 17 private companies and the remaining 35% is publicly owned, through an IPO that closed at the end of 2005 (fig. 10–11).

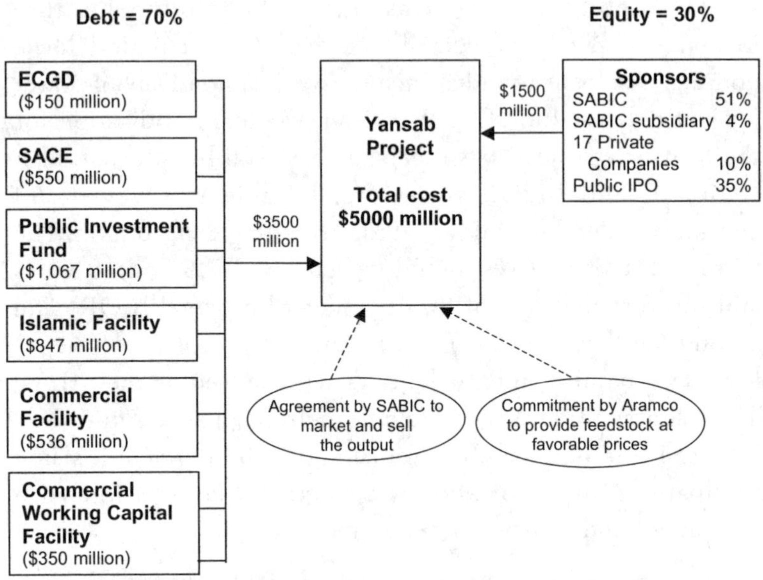

Fig. 10–11. Financing structure of the YANSAB project

The IPO itself was a major success. YANSAB was the first company in Saudi Arabia that could raise such an amount of money at the project inception stage. The IPO was received very well by the market and was oversubscribed by almost 200%.

The project debt was arranged by ABN Amro, which also underwrote the entire debt. The debt is comprised of six tranches: the United Kingdom's and Italy's export finance facilities, ECGD and SACE, provided \$150 million and \$550 million, respectively; \$1,067 million was provided by Public Investment Fund (PIF), which is a Saudi government entity that provides supplementary funding to commercial projects of national interest; and the remaining three tranches were raised on the market. The Islamic facility of \$847 million was one of the largest Islamic issues in the Middle East. A number of financial institutions played significant roles in the syndication process, including ABN Amro, BNP Paribas, Fortis Bank, ING, Mizuho, Standard Chartered, Sumitomo Mitsui, UFJ, Bank of Tokyo-Mitsubishi, Arab Banking Corporation, and Gull International Bank.

Although the debt is mobilized on a nonrecourse basis, the project was perceived to have the support of the government of Saudi Arabia because of SABIC's share in ownership and PIF's financial support (which carried the same terms as the commercial loan and, therefore, no subsidy). In addition, Saudi Aramco has committed to supply YANSAB with favorably priced propane, and SABIC has agreed to market all YANSAB's output and make payments according to a net back formula related to market prices.

Gas T&D project in Bangladesh. Financing of gas T&D investments is still dependent on the government, particularly in poor countries. The Bangladesh example is typical of investment finance in low-income countries.

The project consists of construction of 353 km of gas transmission pipeline for transporting 360 million cubic feet per day (MMCFD) of gas to less-developed parts of the country, a 200 km gas distribution network in the Rajshahi area, and support for gas development and institutional improvements in the gas sector. The project cost is estimated at $413 million. The entire amount is provided by government-guaranteed formal borrowing (fig. 10–12). The government contribution amounts to $178 million and comes from the government's development budget. The official borrowing consists of a $225 million loan from the ADB. In addition, there are two grant components, each amounting to $5 million, from the Norwegian development assistance program and the ADB, to support the project's objectives.

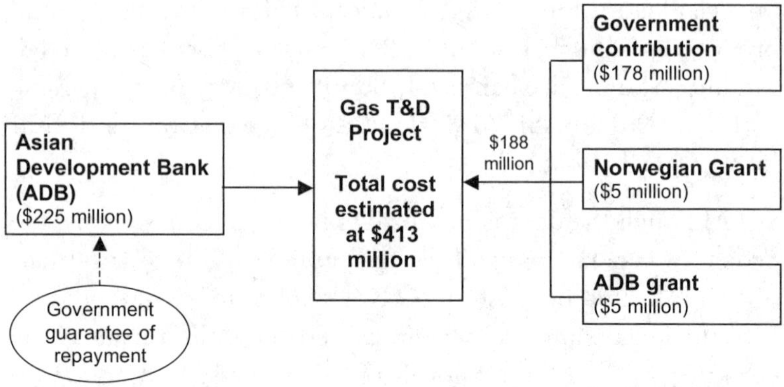

Fig. 10–12. Financing structure of the Bangladesh gas T&D project

Phu My 3 power plant. Financing of the Phu My 3 power plant was an impressive undertaking that demonstrated how Vietnam has moved from public to private finance with government support. The Phu My power complex consists of a series of power plants with a combined capacity of up to 3,600 MW. It is located in the village of Phu My in the southern part of Vietnam. The first phase, of about 1,100 MW, called Phu My 1, was constructed between 1998 and 2000 with funding from the government and Japanese assistance. The second phase, Phu My 2, was implemented through two BOT arrangements, led by Siemens-Mitsui and the EDF. The third phase, Phu My 3, consists of a 715 MW combined-cycle plant with an estimated cost of $412 million; it is based on a BOT arrangement fully owned by BP, which uses gas from Nam Con Son Basin (discovered and developed by BP).

The funding of the Phu My 3 project is based on 25% equity and 75% debt (fig. 10–13). The equity, about $102 million, is contributed by BP. The debt is arranged through three sources: multilateral, bilateral, and commercial. Among the multilaterals, both the ADB and MIGA provide support: the ADB provides a loan of $40 million and a guarantee of $35 million; MIGA provides a guarantee of $40 million. Among the bilaterals, JBIC provides a loan of $99 million, while NEXI provides a guarantee of $95 million. The commercial tranche of $170 million is covered by guarantees from the ADB, MIGA, and NEXI. Commercial lenders are the Bank of Tokyo-Mitsubishi, Credit Agricole Indosuez, Fortis Bank, and Mizuho.

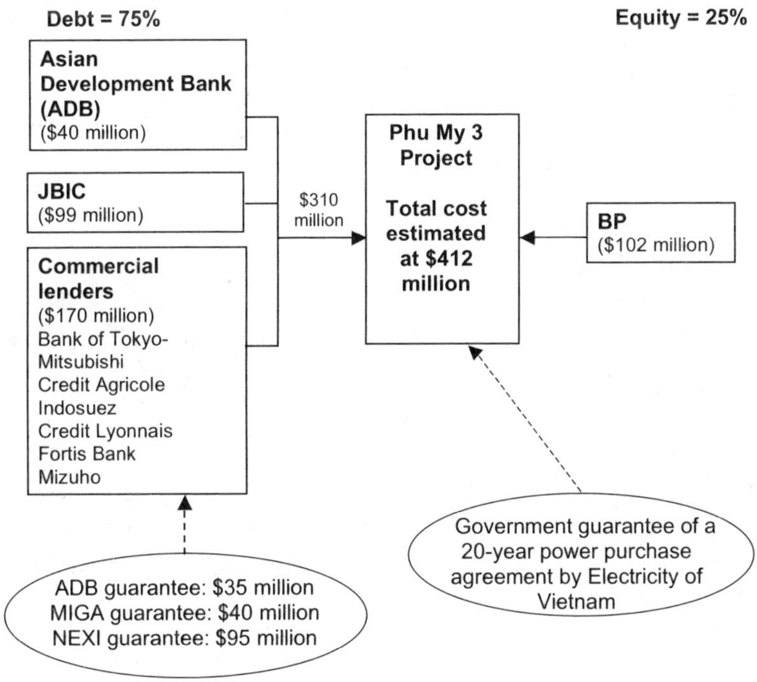

Fig. 10–13. Financing structure of the Phu My 3 power plant

Hidd power and desalination plant. Financing of the Hidd project in Bahrain was arranged in 2006 and has several interesting features. First, a large project of $1.3 billion in a small country was funded on very favorable terms. Second, sponsors avoided contributing any equity. Third, a large portion of the mobilized loans were used for the purchase of a plant through a privatization transaction.

The project entity, Hidd Power Company, is owned by three strong international players: Sumitomo (30%), International Power (40%), and Suez-Tractebel (30%). The company uses the funds that it raises to finance the purchase of an existing power generation and desalination plant (power capacity: 965 MW; desalination capacity: 30 million gallons per day) from the Ministry of Electricity and Water. The project company will use the funds also to construct a new desalination plant (60 million gallons per day).

The funding package consists of a $600 million loan from JBIC and a $390 million loan from a syndicate of commercial lenders (fig. 10–14). These two loans amount to $990 million, representing about 76% of the total project cost. The remaining 24% is expected to come from project sponsors in the form of equity. In this case, the remaining funds will be mobilized through other instruments that do not draw upon the sponsors' resources but are still viewed as equity contributions. The commercial debt of $390 million is arranged by RBS, Mizuho, KfW, SMBC, Gulf International Bank, and Standard Chartered.

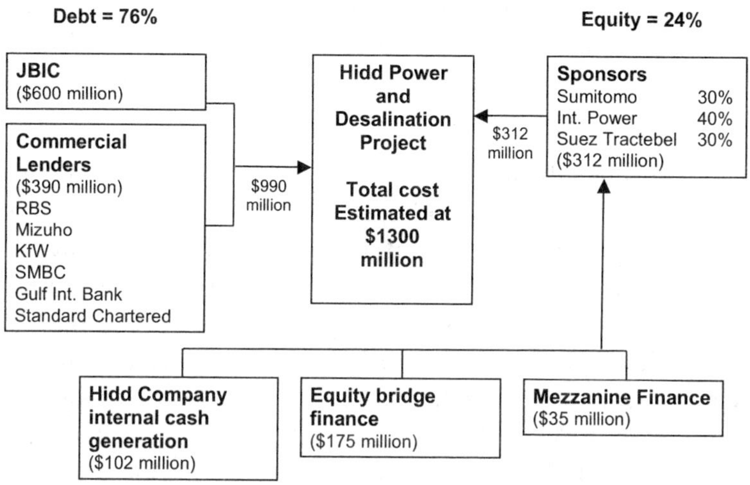

Fig. 10–14. Financing structure of the Hidd power and desalination plant

Bujagali hydropower project. Financing hydroelectric projects in developing countries has been facing serious challenges due to the requirements of large up-front investments and the need for extensive project preparation work in order to handle technical and often controversial environmental issues. Bujagali project in Uganda is an interesting example of a hydropower project that has reached the stage of financial closure despite all the challenges. The project is built by the private sector through a BOT scheme which represents another point of departure from the usual reliance on the state electricity company for construction of hydroelectric projects.

Bujagali project is based on 250 MW capacity, with a small reservoir for daily storage on the Nile River at Dumbbell Island in Uganda. It is expected to generate about 1,438 GWh/year of electricity representing a plant factor of 66%.

The Bujagali project was under preparation for many years; the ownership and financial structures changed a number of times. The financial and security agreements were finalized in 2007 based on 22% equity and 78% debt while the government and the multilateral agencies played a significant role in mobilizing debt and equity resources. The project sponsors (shareholders) include: the Industrial Promotion Service of Kenya (IPS(K)), which is a subsidiary of Aga Khan Fund for Economic Development; Sithe Global - World Power Holding; and the government of Uganda. The shareholders provide $171 million in equity accounting for 22% of the project cost of $798 million. The rest of the project cost will be financed trough borrowing. Multilateral and bilateral agencies provide $512 million and commercial lenders provide $115 million.

Bujagali financial structure represents a case where many softer sources of finance have been mobilized and utilized to form a BOT scheme. The same circle of support is utilized to bring comfort and manage risks in project implementation and operation. Figure shows the contractual agreements among various parties. The government has awarded a concession to the project company. A state entity commits to purchase the project output while this commitment is further guaranteed by the government. The commercial lenders are guaranteed by IDA, which itself receives a counterguarantee from the government. Most other risks are mitigated through the standard practice in BOT projects.

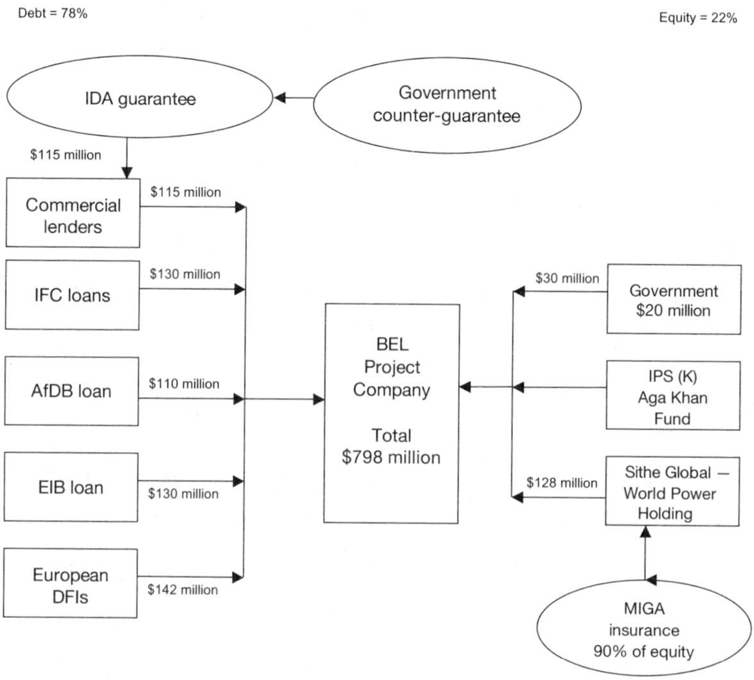

Fig. 10–15. Financing Bujagali Hydropower Project

Financing Nam Theun II (NT2) Hydropower Project. Financing of the NT2 project in Laos has set a new milestone in formulating hydroelectric projects (fig. 10–15). The project cost is estimated at $1,350 million. This is a very large project to be implemented in a small country with rather low level of creditworthiness. There are also significant environmental issues to be handled. There are two important factors enabling this project reach financial closure. First, more than 90% of the project output will be exported to Thailand where the purchaser of the electric power is quite credible and stable. Second, the impact of NT2 on the Lao economy is substantial; therefore, the project receives a lot of support from many donors.

NT2 project involves a hydropower facility with an installed capacity of 1,070 MW providing 995 MW of power to export to Thailand and an additional 75 MW for domestic use. The hydropower facility comprises a dam 48 meters high and a 450 km^2 reservoir on the Nam Theun River in Laos. The project also includes the associated transmission systems from the powerhouse to the Thai grid and the Lao grid.

A project company—Nam Theun 2 Power Company Ltd. (NTPC)—has been formed (fig. 10–16). It has received a concession from the Lao Government to build the project on a build-own-operate-transfer (BOOT) basis. The concession period is 25 years. The concession allows NTCP to sell electricity to the Electricity Generating Authority of Thailand (EGAT) and Electricité du Laos (EDL). Subsequently, NTPC has signed power purchase agreements (PPA) with EGAT and EDL on a take-or-pay basis for 995 MW and 75 MW, respectively.

There are four project sponsors (shareholders) in NTCP, including Electricité de France International (EDFI); Italian-Thai Development Public Company Ltd (ITD), Thailand; Electricity Generating Public Company Ltd (EGCO), Thailand; and Lao Holding State Enterprise (LHSE) - a newly created enterprise to hold the Lao Government's equity in NTPC. Each shareholder is contributing $87.5 million. The LHSE portion is provided by the government which itself is funding it through a $20 million IDA grant, $6.2 million AFD grant, $41 million EIB loan, $12.5 ADB loan and part of the prepayment from NTPC to the government for the project concession fees.

The debt portion of the financing plan amounts to $1 billion and is secured by NTPC from a combination of multilateral, bilateral and commercial sources. Multilaterals and bilaterals provide direct loans to NTPC, and also guarantee the loans from international commercial banks.

This project has been under preparation for many years. Its financing and ownership schemes have been revised numerous times. The final financial package reached closure in 2006.

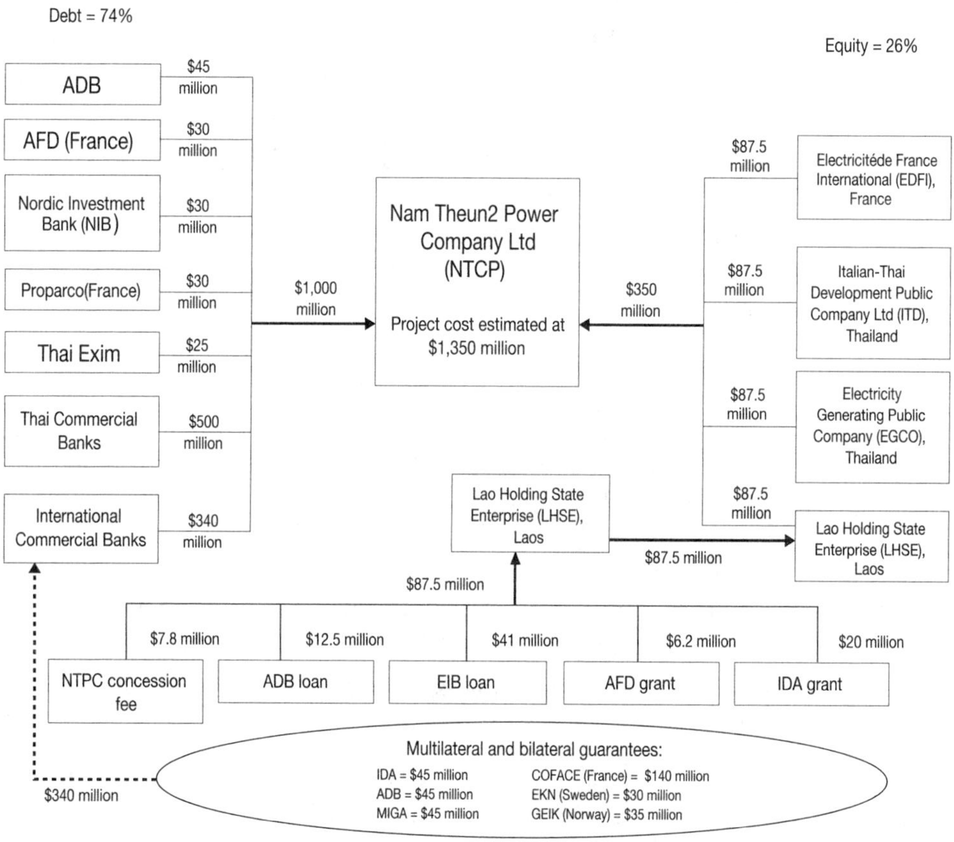

Fig. 10–16. Financing structure of the Nam Theun II Hydropower Project

Wind energy projects in Brazil. Financing of renewable energy projects has been rather difficult in most developing countries. However, in recent years, the average cost of power generation for some of the renewable energy technologies has declined. Also, public awareness and financial incentives for protecting the local and global environment have added to the attractiveness of renewable energy, at least in some developing countries.

Brazil, which is well known for its significant hydroelectric power potential, has in recent years recognized wind power as another source of renewable energy that should be tapped for electricity generation. Development of wind power would, however, need various types of support at its early stages. Government support for wind power translates into the following:

- Offering long-term contracts to investors to buy their output at attractive prices
- Providing loans through the Brazilian National Social and Economic Development Bank (BNDES)

The first large-scale wind power farm will be implemented by Energias Renovaveis do Brasil Ltd. (Enerbrasil), which is a Brazilian wholly owned subsidiary of Iberdrola SA of Spain, one of the leading companies worldwide in the renewable energy business.

The Enerbrasil wind project encompasses the construction, operation, and maintenance of a 49 MW wind power park located in the municipality of Rio do Fogo, in the northeast of Brazil. The project cost is estimated at $92.6 million and will be funded through 30% equity and 70% debt (fig. 10–17). Equity will be provided by the project sponsor, while the IFC will also make a $5.5 million investment in the project. Debt is provided through a loan from BNDES.

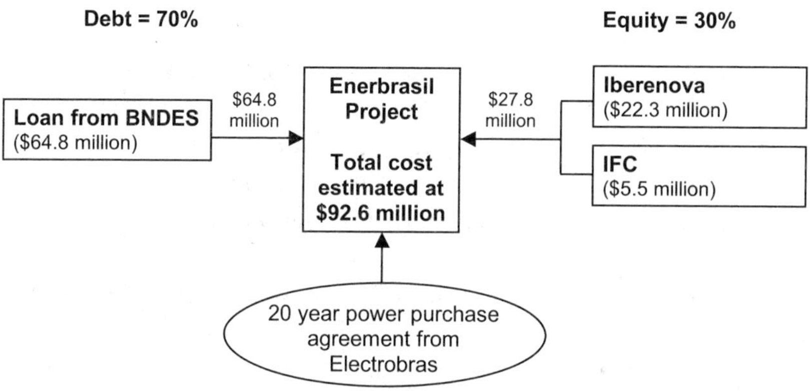

Fig. 10–17. Financing structure of the Enerbrasil wind power project

Rockfort power project. This Jamaican project is an interesting and well-known case of project finance, owing to its creative financial structure for a small project in a low-income country. The project is also noteworthy because of its extensive use of the bond market and tax-exempt financing to reduce the cost of equity and debt to the project.

The project consists of a 60 MW diesel power plant in Rockfort. The project cost is $144 million, which includes the base cost ($120 million), as well as standby facilities ($24 million).

Project sponsors include affiliates of Hydra-Co Enterprises Incorporated (at the time, a subsidiary of a utility in New York State), U.S. Energy Corporation, Precursors Systems Incorporated, and International Energy Finance (IEF). This group of private power developers formed a special-purpose project company in Jamaica.

The financing structure was based on 30% equity and 70% debt (fig. 10–18). The four sponsors committed $16 million in equity. The remaining equity was committed by the United Kingdom's CDC ($7 million), UtilCo Group Incorporated (a subsidiary of a U.S. investor-owned gas and electric utility; $12 million), and the Energy Investors Fund (a private investment fund in the United States; $8 million). The debt is financed by CDC ($20 million) and by the bonds of the Caribbean Basin Projects Financing Authority (CARIFA), with a five-year maturity and which are supported by a consortium of international banks through a $83 million letter of credit issued by the Deutsche Bank. Following project completion and at the maturity of the CARIFA bonds, term financing would be provided by the Jamaica Private Sector Energy Fund (PSEF; $81 million). PSEF is a government-owned lending facility financed by the World Bank and the IDB, with each providing $40.5 million. Project sponsors also secured a $50 million guarantee from MIGA to insure equity and debt against political risk.

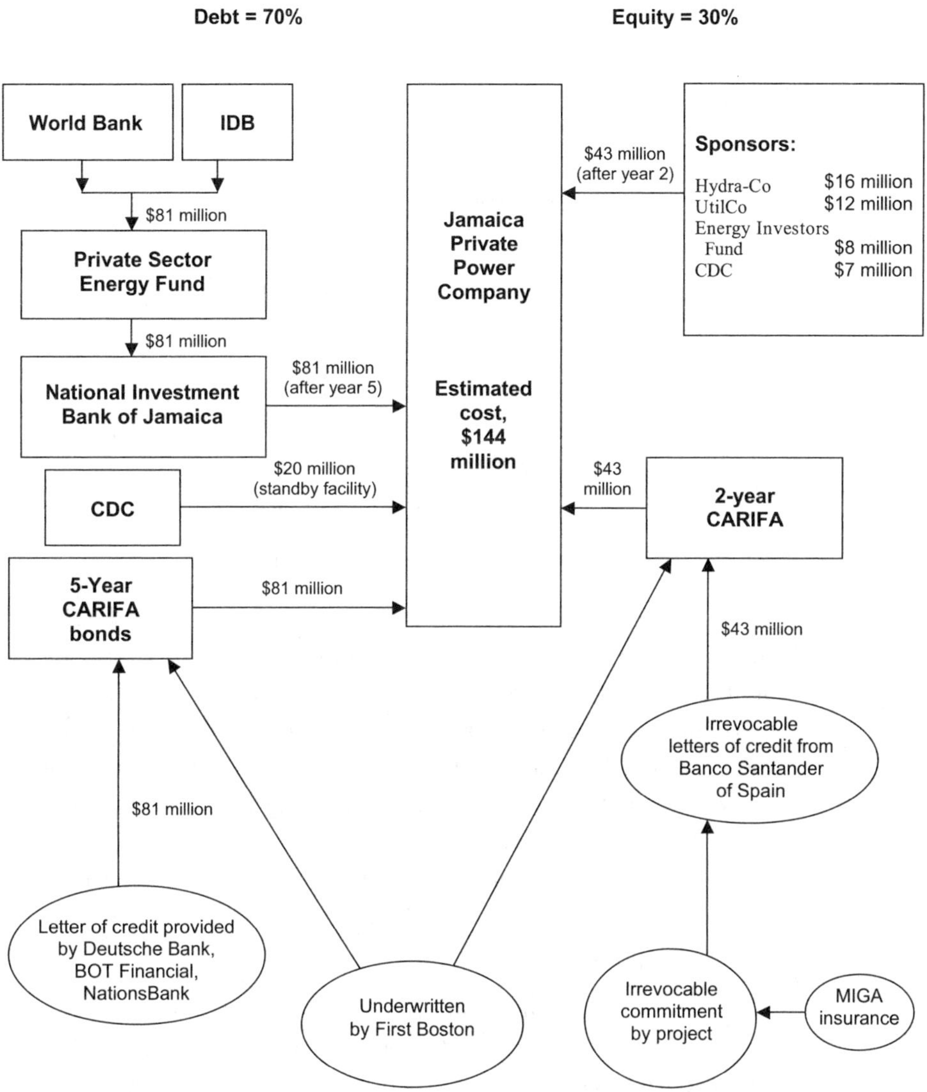

Fig. 10–18. Financing structure of the Rockfort power project

Thus, the apparent financing structure is as follows:

Equity (30%):	
Hydra-Co	$16 million
UtilCo	$12 million
Energy Investors Fund	$8 million
CDC	$7 million
Subtotal	$43 million
Debt (70%):	
PSEF	$81 million
CDC	$20 million
Subtotal	$101 million
Total	$144 million

In the preceding financing structure, except for CDC, none of the aforementioned parties put any funds, direct or apparent, into the project. Instead, they raised $122 million by issuing bonds, also referred to as *936 funds.* The title of the fund relates to a section in the U.S. tax code that provides for a tax exemption on earnings of U.S. corporations in Puerto Rico, provided that these earnings are reinvested in Puerto Rico or certain Caribbean countries (including Jamaica). Because of the reinvestment obligation, these funds carry low interest rates (below LIBOR).

Sponsors of this Jamaican private power project raised $122 million through CARIFA, the Puerto Rican financial institution set up to channel investment funds from U.S. corporations to private sector projects in the Caribbean. CARIFA offered three bond issues—two totaling $43.2 million, to fund the equity, and one for $81 million, to fund the debt. The last of these bond issues was supported, as mentioned earlier, by a letter of credit from a commercial bank rated AAA. The equity portion of the issue has a maturity of two years and is indirectly backed by commitment from the equity investors through a letter of credit issued by Banco Santander. The debt portion of the issue has a maturity of five years and is indirectly backed by the PSEF (which is supported by the World Bank and the IDB). The five-year bond issue was rated AAA, and the two-year bond issue was rated AA−, by S&P, based on the strength of the supporting letters of credit.

The successful mobilization of financing was attributable to the following factors:

- Securing of a 20-year power purchase agreement with the state-owned power utility. The agreement was further guaranteed by the government.
- Signing of an implementation agreement between the project company and the government, which set the general parameters for the company's investment in power generation.
- Signing of a 20-year fuel supply agreement with the government-owned Petrojam Refinery.
- Support of international commercial banks, which issued letters of credit based largely on the project sponsors' guarantee of repayment of equity bonds, and the World Bank and IDB funding agreements made through PSEF.
- Underwriting of both bond issues by First Boston.
- Guarantee by MIGA against political risks.

Power T&D projects in Argentina, Kazakhstan, and India. Power transmission investments are normally supported by the government and implemented by the national power company. Accordingly, the funding of such projects comes from the power company's internal cash generation, the government's development budget, and official borrowing from multilateral and bilateral lenders. There are nevertheless attempts to diversify the source of funding either through nonsovereign borrowing by the electricity company or through private sector sponsorship of specific segments of the transmission system.

The case of Argentina provides a typical example of transmission investments funded by the government, while the case of Kazakhstan demonstrates a move by the electricity company to diversify its sources of borrowing. The case of India focuses on a specific transmission project that is constructed by the private sector.

The Argentinean power transmission project Norte Grande Electricity Transmission Program commenced implementation in 2006. The program consists of an extrahigh-voltage system that ties the northeast and the northwest sides of the country and corresponding medium- and high-voltage reinforcement systems in each of these two regions. The total cost of the project is estimated at $725 million. The financing plan (fig. 10–19A) is based on a large loan, of $580 million, from the IDB, which is provided on an official basis—that is, with the guarantee of the government of Argentina. The remaining $145 million is provided from the state's resources.

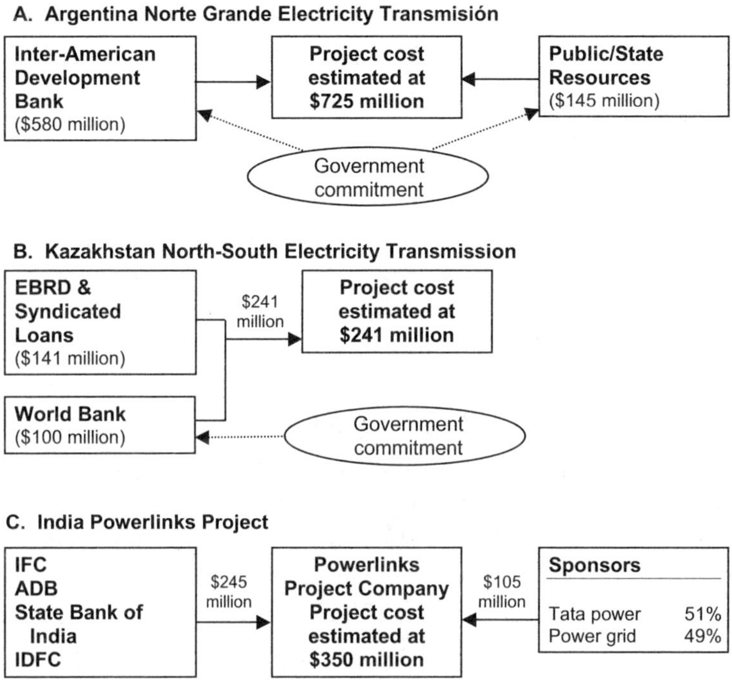

Fig. 10–19. Financing structures of transmission projects in Argentina, Kazakhstan, and India

The North-South Electricity Transmission Project in Kazakhstan aims at the construction of a 1,115 km, 500-kilovolt system and corresponding facilities to transport power from generating units in the north to the major load center in the south of the country. The project cost for the final two phases is estimated at $241 million, which is all funded from borrowing (fig. 10–19B). However, the debt consists of

- A nonsovereign loan of $53 million from the EBRD
- An EBRD-led syndicated loan of $35 million

- A $53 million loan from the Development Bank of Kazakhstan
- A $100 million sovereign-guaranteed loan from the World Bank

The Powerlinks Project in India is a public-private joint venture of Tata Power, which is a prominent private power company in India, and the Power Grid Corporation of India, which is the country's national transmission company. Powerlinks has been established as an independent entity with shareholdings of 51% by Tata Power and 49% by the Power Grid Corporation, with the objective of constructing a power transmission system from Siliguri, in West Bengal, to Mandola, near Delhi, on a BOT basis. The transmission system will have a capacity of 3,000 MW and will carry power from the eastern region to the power-deficient northern region of India. The project cost is estimated at $350 million. The funding is based on 30% equity and 70% debt (fig. 10–19C). The debt portion is provided by the project sponsors in proportion to their shares in the company. The debt is financed on a nonrecourse basis from the IFC, the ADB, the State Bank of India, and the Infrastructure Development Finance Company (IDFC, a private infrastructure fund).

Power distribution projects. Power distribution investments are mostly financed by the public sector. Still, numerous attempts have been made to transfer the responsibility to the private sector.

The case of Power Distribution Reconstruction Project in Bosnia and Herzegovina demonstrates a situation in which there is no alternative but public sector finance. The project proposes to reinforce the distribution facilities of the three regional utilities, to improve the reliability and efficiency of electricity distribution. The project cost is estimated at $78 million. Funding includes a $65 million loan from the EBRD; the remaining $13 million is provided through public resources. The EBRD loan is made to the government, who then onlends it to the three utilities.

11 Environmental And Social Concerns

Since the 1980s, growing environmental awareness has been accompanied by a more focused attention on the interactions between the natural environment and economic development. Although this attention has to some extent been translated into development policies, the principle strategy for minimizing damage to the environment has been the setting of environmental standards for construction and operation of commercial facilities—in particular, energy-producing and -processing units. These standards are imposed not only on new plants but also, sometimes, on plants already in operation. Indeed, application of stringent environmental standards to existing plants has caused much concern and, in some cases, tremendous financial loss.

Concerns about the environmental impacts of a project fall into three categories:

- Legal issues. Most developing countries have a legal requirement to assess the environmental impacts of energy projects and to ensure that standards are met during construction and operation of a plant.
- Possible financial damages. Experience has now demonstrated that returns from an investment—and even

the investment itself—can be ruined if an energy plant does not adhere to environmental standards.

- Social responsibilities involving the local, regional, and global environments. These are much broader concerns raised by certain government agencies, bilateral and multilateral development agencies, and various environmental groups, as well as neighboring communities that may be affected.

The legal requirement to assess the environmental impact of any industrial project was introduced in the United States in 1970, when the National Environmental Policy Act (NEPA) was signed into law by President Nixon. Today, most developing countries have passed similar laws, although details and enforcement vary considerably.

The task of assessing environmental impacts of a project has become increasingly complex, as the framework of analysis has been broadened over time. When environmental assessment was conceived, it was regarded as an add-on component of cost-benefit analysis, intended to incorporate the potential environmental damages into the cost of the project and to select the least-cost remedial actions. Soon, it became necessary to expand the framework to cover the direct and indirect impacts on climate and air quality, geology and soils, hydrology, and so on. It was further accepted that environmental assessments should be carried out by multidisciplinary teams able to examine social impacts. In association with the preceding extension, it became necessary to establish a link between environmental assessment and policy planning and to work with various interested and affected parties to ensure that all relevant views would be taken into account. Project sponsors are often asked to go beyond the scope of the project and consider the environmental impacts of the entire fuel chain or to examine how remedial actions can be combined with other policies (e.g., energy efficiency improvement strategies).

Especially since the Rio Conference in 1992, examination of project impacts on the regional and global environment, although not typically a requirement of the host government, is demanded by some financiers.

Clearly, the aforementioned extensions to the examination of environmental aspects of projects have resulted in a complex, time-consuming, and sometimes unmanageable process. At the same time, however, the state of the art has progressed significantly, particularly with regard to communication with interested parties. Project sponsors are expected to be responsive to most if not all of these aspects at appropriate stages of project formulation, construction, and operation.

It is also useful to understand the concept of *sustainable development* as practiced by many financiers. Sustainable development encourages project sponsors to take a holistic approach to project design. The traditional practice is to design a project based on technical and economic parameters while ensuring that social and environmental requirements are met. The holistic approach advocates the view that environmental and social aspects should be treated not as add-ons but as the main components of project design, to ensure sustainable development. This is depicted in figure 11–1, in which economic, sociological, and ecological objectives are viewed as parallel dimensions in project design.

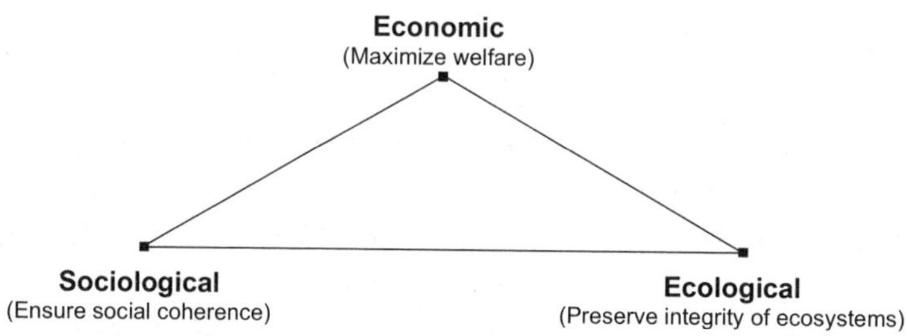

Fig. 11–1. Framework for sustainable development

While the framework depicted in figure 11–1 may appear to be only philosophical, it nevertheless results in some practical procedures. First, the project design should arrive at a solution that results from a consensus—or, more often, a compromise of the economic, sociological, and ecological considerations. That is, in practice, the project design emerges through negotiation of these considerations and examination of viable alternatives. Second, the concept of sustainable development has been operationalized through a set of safeguard policies. As the leading agency in this area, the World Bank has identified 10 key policies that are essential in ensuring that potential adverse environmental and social consequences are identified, minimized, and mitigated:

1. Environmental assessment (EA)
2. Involuntary resettlement
3. Forests
4. Safety of dams
5. Projects on international waterways
6. Projects in disputed areas
7. Indigenous peoples
8. Natural habitats
9. Pest management
10. Physical cultural resources

The details of these policies are described on the World Bank Web site. Not all of these policies are triggered for every project. Depending on the features and location of the project, one has to examine the applicability of the safeguard policies. The first policy (preparation of the EA) is applicable to the majority of energy projects. Among the other policies, involuntary resettlement is more frequently triggered in energy projects, in which there is a need to acquire land for a project site or access road right-of-way (ROW), and so on. Construction of hydroelectric projects would normally trigger the majority of these safeguard policies. However, the EA encompasses a review of all applicable safeguards.

Summary 11–1. Environmental Issues in Project Analysis

- Financiers insist that project sponsors address relevant environmental impacts sufficiently, because environmental problems can result in serious financial losses.
- Financiers have become increasingly committed to prevention of environmental damages that result from energy projects and set strict guidelines for review of the environmental impacts of such projects.
- Assessment of the environmental impacts of oil, gas, and power projects has become broader and more complex, covering many direct and indirect impacts—including impacts on climate and air quality, geology and soils, hydrology, social and economic structures, and regional and global environments.
- The concept of sustainable development is increasingly used to indicate the need for a holistic approach to project design. This approach has been put into practice at the World Bank through the introduction of 10 safeguard policies.

The Overall Framework of the EA

EAs are required for the majority of energy projects and by all financiers. Most financiers need a complete EA prior to project appraisal. The EA should be presented concisely and effectively to provide assurance that all important environmental risks have been identified and that a comprehensive plan has been formulated to mitigate and manage them. The World Bank Group has issued a set of guidelines for preparation of the EA. Most other financiers accept these guidelines. The World Bank guidelines for preparation of environmental assessment are summarized in appendix B.

The essence of the EA is the prediction of alternative future states of resources and environment that will result from various development paths, including no development at all. This is achieved by making the following assessments:

- *Measuring the environmental quality parameters before the project has begun*. The quality of air and water supplies, the fertility of soil, and the nature and quality of habitat must be measured. These baseline data should also include natural changes that might occur in the absence of the proposed project.
- *Identifying significant impacts that the construction and operation of a plant would have on the environment.* The World Energy Conservation Strategy suggests consideration of at least three criteria: first, the period and the geographical area over which the effect will be felt (including the number of people affected and the resources that would degrade); second, the urgency (i.e., how quickly a natural system might deteriorate and how much time is available for its

stabilization or enhancement); and third, the degree of irreversibility of damages to air, land, water, quality of life, plants, and animals.

- *Quantifying the environmental impacts*. The objective is to evaluate the environmental changes induced by a project that affect human health and welfare in either the short or the long term. This includes measurement of eventual damages to human health, agricultural productivity, and other biophysical and socioeconomic changes likely to result from the project. Some of these damages may be measured on the basis of market prices; others may be assessed in relation to international benchmarks and other proxies. However, still other impacts remain nonquantifiable; these are analyzed qualitatively. Qualitative and quantitative assessments of environmental impacts of a project incorporate the following aspects: First, assessment includes both positive and negative impacts of a project. Second, benefits and costs display a useful symmetry—a benefit forgone is a cost, and a cost avoided is a benefit. Third, assessment of a project is made on a "with and without" basis, extended to alternative designs, sites, technologies, and modes of operation. Fourth, assessment includes costs and benefits of mitigation measures, including those reasonably proposed by concerned constituencies.

Summary 11–2. Environmental Assessments

- Most financiers require an environmental assessment (EA) for energy projects.
- The EA examines (1) the baseline conditions, (2) the impacts of project construction and

operation on these conditions, and (3) alternative development paths if the project is not constructed.

- Financiers have specific guidelines for preparation of EAs. The guidelines used by the World Bank Group are normally accepted by other financiers. These guidelines are summarized in appendix B.

The remainder of this chapter addresses specific environmental concerns that project sponsors should consider in preparing EAs for oil, gas, and power projects. Also, the environmental standards that are usually used in EAs to demonstrate that environmental risks of the project are kept below harmful levels are summarized.

The Environmental Impacts of Energy Projects

Oil and Gas Development

Oil and gas production requires extensive on-site industrial activities over the life of the field. Construction of well sites, access roads, airfields, gathering and transport pipelines, and ancillary support facilities results in extensive surface disturbance, construction traffic, noise and air emissions, and influx of construction crews. Continual noise and air emissions are a part of equipment operation, as is discharge of treated formation waters and oil spills, which may affect soil and groundwater quality. Air pollution also results from flaring of unwanted gases, sour-gas (hydrogen sulfide) discharge, and burning of oil waste pits. Potential catastrophic accidents include well blowouts, with uncontrolled oil and gas release, and fires in refineries or gas-processing plants.

Offshore drilling and production facilities, vessel traffic, and coastal pipeline landings may interfere with fishing and with pleasure-boat use of the coastal area. Noise from overflights, near-shore drilling, production operations, port traffic, and processing plant operations will be distracting. Offshore and onshore facilities have visual impacts as well. The initial construction force tends to be transient and is soon replaced by the operations staff, usually smaller and permanent.

Catastrophic oil spills from well blowouts, pipeline ruptures, or shipping accidents threaten marine mammals, sea- and shorebirds, and coastal areas. Control and cleanup of a major oil spill, well blowout, or fire—with emergency deployment of large crews, equipment, and supplies—creates a severe but temporary disruption of other activities in the coastal area. Residual effects from a spill can include oil-stained beaches, boats, and shore facilities and migration of certain marine species, at least temporarily.

Oil and gas pipelines

The major facilities associated with oil or gas pipelines include the pipeline itself; access or maintenance roads; the receiving, dispatch, and control station; digging stations; odorizing stations; and the compressor or pump stations. Because of internal friction and changes in elevation encountered along the line, booster stations are required at regular intervals for long-distance crude oil and product pipelines. For gas transmission lines, compression stations are installed at appropriate intervals along gas transmission lines to maintain pipeline pressures. Pipelines may transport unrefined oil or gas from a wellhead to transfer or processing facilities. Refined oil and gas may be transported by pipeline to an end user, such as a petrochemical plant or a power plant, or to an export terminal.

Offshore, near-shore, and upland oil and gas pipelines have different environmental impacts. The magnitude of the impacts depends on the type and size of pipeline installed, and the significance depends on the degree to which natural and social resources are affected.

In the case of offshore pipelines, areas of concern include the following:

- The trenching and turbidity associated with pipelaying in offshore and near-shore areas may disrupt bottom-feeding organisms.
- Construction of the pipeline can temporarily resuspend bottom sediments. The subsequent resettlement of sediment may alter aquatic habitats and change the composition of species. The significance of these effects will depend on the type and importance of the aquatic organisms affected. For example, the alteration of sea grass beds or coral reefs, which are considered important feeding and breeding habitats for fish and other animals, may be greater in consequence than the alteration of deep offshore benthic habitats.
- If pipeline trenching occurs in near-shore and offshore areas where toxic chemicals have accumulated in the sediments, the laying of the pipeline can result in a resuspension of these toxic sediments and can temporarily lower water

quality immediately above the pipeline. Bioaccumulation of the toxic chemicals may occur in aquatic organisms (e.g., fish and shellfish).

- In near-shore and offshore areas used for bottom fishing, pipelines can interfere with bottom trawling, resulting in loss or damage to fishing equipment, as well as accidental ruptures to pipelines. Anchor dragging can also result in pipeline damage and oil spills.

For onshore pipelines, environmental damages can occur in the following areas:

- Installation of pipelines can promote erosion. In hilly areas, this can lead to instability and landslides. Runoff and sedimentation can lower water quality in rivers and streams during construction.
- Installation of pipelines and maintenance of roads can alter drainage patterns, blocking water flow and raising the water table on the upslope side of the pipelines, and can kill or reduce vegetation. If a pipeline cuts through a large forested area, the impact may be significant. Water supply to wetlands can be altered.
- Creation of ROWs can introduce exotic plants able to replace native vegetation. In addition, pipeline installation can fragment the habitat of natural areas (e.g., wildlands), resulting in the loss of species.
- In developed areas, oil and gas pipelines and substations can displace inhabitants or otherwise compromise land use. Some types of agricultural activities may be affected during construction.

- Aboveground pipelines can create barriers for humans and migratory wildlife. The significance of the effect depends on the length and location of the pipeline.
- Archeological sites are vulnerable to damage or loss during pipeline construction.
- Pipeline construction can temporarily interrupt traffic. This may be significant in developed areas if the pipeline crosses major transportation routes.
- Ruptures and leaks, as well as wastes generated at pump and transfer stations, can contaminate soils, surface water, and groundwater. The significance of this contamination depends on the type and size of the leak, the type and volume of wastes generated, and the degree to which the natural resource is affected. Ruptures of oil pipelines crossing rivers and other bodies of water or wetlands can result in significant environmental damage.
- Leakage or rupture of gas pipelines can cause explosions or fires.

Petroleum refining

The environmental impacts of petroleum refining include gaseous emissions, wastewater discharges, solid wastes, noise, odor, visual or aesthetic effects, and soil contamination. Atmospheric emissions include particulates, hydrocarbons, carbon monoxide, sulfur oxides, and nitrogen oxides. They emanate from various sources, including catalytic cracking units, sulfur-recovery processes, heaters, vents, flares, and product or raw material storage. Pump seals and valves can be sources of fugitive emissions. The combination of emissions can produce noxious odors.

The major pollutants present in petroleum refinery wastewater discharges are oil and grease, ammonia, phenolic compounds, sulfides, and organic acids, along with chromium and other metals. These pollutants may be expressed in terms of 5-day biochemical oxygen demand (BOD_5), chemical oxygen demand (COD), and total organic carbon (TOC). In addition, there is potential for serious surface-water, soil, and groundwater contamination from leaks or spills of raw materials or products. Cooling-water blowdown, flushing and cleaning water, stormwater runoff, and percolation from tank frames, pipe racks, product-loading areas, and processing areas can also degrade surface water and groundwater.

Refineries generate large volumes of solid wastes. Chief among them are catalytic fines from cracking units, coke fines, iron sulfides, filtering media, and various sludges (from tank cleaning, oil and water separators, and wastewater treatment systems).

Petroleum refining can be a noisy operation. Sources of noise include high-speed compressors, control valves, piping systems, turbines and motors, flares, air-cooled heat exchangers, fans, cooling towers, and vents. Typical noise levels range from 60 to 110 decibels (dB) at a distance of one meter from the source.

A major release or spill of raw materials, products, or wastes can be environmentally catastrophic, especially to marine or aquatic ecosystems. As in any operation that involves hydrocarbons, fire and explosion are constant possibilities.

Thermal power projects

Thermal power projects include gas-, oil-, and coal-fired conventional plants, as well as combined-cycle, gas turbine, and diesel power plants. The major components of thermoelectric projects include the power system (i.e., power source, turbine, and generator) and associated facilities, which may include the cooling system, stack-gas–cleaning equipment, fuel storage and handling areas, fuel delivery systems, solid waste storage areas, worker colonies, electrical substations, and transmission lines. Environmental harm can occur during both construction and operation of these plants. Construction impacts are caused primarily by the following site preparation activities: cleaning; excavation; earthmoving; dewatering, dredging, or impounding streams and other water bodies; establishing laydown areas; and developing borrow and fill areas. The large number of workers employed in the construction of power plants can have significant sociocultural impacts on local communities.

The environmental impacts during plant operation include

- The impact of air emissions on local, regional, and global environments
- The impact of cooling water and waste heat on aquatic plant and animal communities
- The impact on the local community

Air emissions. Thermal plants are considered major air emission sources that can affect local and regional air quality. Sulfur dioxide (SO_2), oxides of nitrogen (NO_x), carbon monoxide (CO), carbon dioxide (CO_2), and particulates (which may contain trace metals) are emitted from the combustion of fuels by thermoelectric projects. The amounts of each depend on the type and size of facility, the type and quality of fuel, and the manner in which fuel is burned. Dispersion and ground-level concentrations of emissions are determined by a complex interaction among the physical characteristics of the plant stack, the physicochemical characteristics of emissions, the meteorological conditions at or near the site when emissions travel from the stack to the ground-level receptor, the topographical conditions of the plant site and surrounding areas, and the nature of the receptor (e.g., people, crops, and native vegetation).

Emissions from thermoelectric projects can act as precursors of acid precipitation, particularly when the fuel is coal, which has a high sulfur content. Acid precipitation accelerates deterioration of buildings and monuments, radically alters aquatic ecosystems, and damages vegetation. Combustion of fossil fuel in thermoelectric plants also generates CO_2 and NO_x; global warming has been attributed to increased levels of CO_2 and NO_x in the atmosphere.

Cooling water and waste heat. Many power generating plants that use steam also have once-through cooling systems. If the high volume of water used by large plants with these cooling systems is taken from natural water bodies, such as rivers or bays, aquatic organisms may be entrapped in the cooling system. This can significantly reduce populations of fish and shellfish, some of which may be commercially important.

Discharges of heated water can elevate ambient water temperatures, which can radically alter existing aquatic plant and animal communities, favoring organisms that are suited to higher temperatures. These new communities are then vulnerable to the opposite effect—sharp reductions in ambient water temperature—when plants shut down because of failure or scheduled maintenance.

Use of evaporative cooling towers reduces the volume of water that must be withdrawn for cooling to that needed to offset evaporation. Towers eliminate thermal discharge but produce cooling-water blowdown, which must be discharged. In colder climates, a beneficial alternative is to reduce the temperature of cooling-water discharge by using waste heat to make hot water or steam for the heating of buildings or aquaculture ponds. Either form of cooling entails some loss of water, reducing the volume available for drinking, irrigation, navigation, and other uses in water-short areas.

Impacts on the community. One of the major impacts from power plants involves the influx of workers to build the plant. Several thousand workers may be required during the several years of construction of a large plant, and several hundred workers will be needed thereafter for its operation. There is potential for great stress where the host community is small. A *boomtown* effect, or condition of induced development, can result. This strains the existing community infrastructure—schools, police and fire protection, and medical facilities, among others. Similarly, the influx of workers will change local demographic patterns and disrupt local social and cultural values, as well as the living patterns of the residents. Another potential impact is the displacement of the local population because of land requirements for the plant site and associated facilities. Significant disruption of local traffic can occur from the construction and operation of a thermoelectric plant. In addition, large power plants can be visually obtrusive and noisy.

Hydroelectric power projects

Hydroelectric projects include dams, reservoirs, canals, penstocks, powerhouses, and switchyards for the generation of electricity. The dam and reservoir may be multipurpose; if watershed rainfall, stream-flow characteristics, and water and power usage patterns permit, then hydroelectric reservoirs can also provide one or more of the following services: irrigation, flood control, water supply, recreation, fisheries, navigation, sediment control, ice-jam control, and glacial-lake outburst control. However, these are competing uses for water stored behind dams, and each may imply a different diurnal or annual operating rule curve for the reservoir.

The principal sources of impacts in a hydroelectric project are the construction and operation of a dam and reservoir. Large dam projects cause major environmental changes over wide geographical areas.

The area of influence of a dam and its reservoir extends from the upper limits of the reservoir to as far downstream as the estuary, coastal, and offshore zones and includes the reservoir, dam, and river valley below the dam. Although direct environmental impacts are associated with the construction of the dam (e.g., dust, erosion, borrow, and disposal problems), the greatest impacts result from the impoundment of water, which floods land to form the reservoir and alters the flow of water downstream. These effects have direct impacts on soils, vegetation, wildlife and wildlands, fisheries, climate, and human populations in the area.

The dam's indirect effects include those associated with the building, maintenance, and functioning of the dam (e.g., access roads, construction camps, and power transmission lines) and the development of agricultural, industrial, or municipal activities made possible by the dam.

Environmental aspects of hydropower projects have become increasingly difficult to address satisfactorily. Indeed, some critics claim that the environmental costs of dams outweigh their benefits and that the construction of large dams is unjustifiable. Others contend that in some cases, environmental costs can be avoided or reduced to an acceptable level by implementing cost-effective corrective measures. Although for at least some hydroelectric power projects, the environmental impacts are undoubtedly manageable, project sponsors still must be prepared to go through a long and complex process of convincing financiers that the environmental risks are under control.

Power transmission projects

Electric power transmission lines affect the environment during construction, operation, and maintenance. Clearing of vegetation from sites and ROWs, along with construction of access roads, tower pads, and substations, are the primary sources of construction-related impacts. The operation and maintenance of the transmission line involves chemical or mechanical control of vegetation in the ROW and occasional line repair and maintenance. These activities, plus the physical presence of the line itself, can affect the environment.

The effects of short transmission lines may be localized; however, long transmission lines may have regional effects. In general, the environmental impacts on natural, social, and cultural resources increase with line length. The magnitude and significance of the impacts also increase as the voltage of the lines increases and requires larger supporting structures and ROWs. Operational impacts also increase with voltage. For example, electromagnetic field (EMF) effects are significantly greater for 1,000-kilovolt lines than for 69-kilovolt lines.

On the positive side, power line ROWs, when properly managed, can be beneficial to wildlife. Cleared areas can provide feeding and nesting sites for birds and mammals. Power lines and structures can also serve as nesting sites and perches for many birds, especially raptors. Moreover, the increased habitat diversity that results at the contact between the right-of-way and the existing vegetation is well documented in biological literature as the *edge effect*.

Summary 11–3. Environmental Impacts of Power Generation Projects

- Power generation projects are particularly subject to pronounced environmental concerns.
- Environmental concerns are raised about all types of power generation (coal, oil, gas, hydroelectric, and nuclear); however, gas-based power plants are considered least harmful.
- Environmental impacts can occur during plant construction and operation.
- Construction impacts of thermal power projects follow site preparation because of the large numbers of workers employed to build the plant and the relocation of residents from the plant vicinity.
- Environmental impacts during plant operation include hazardous air emissions, effects of cooling water and waste heat on aquatic plant and animal communities, and socioeconomic impacts on local communities.
- The main impact of hydroelectric projects occurs during construction of the dam and reservoir and stems from impoundment of water, flooding of land, and alteration of water flow downstream, which irreversibly affects soils, vegetation, wildlife and wildlands, fisheries, climate, and human populations in the area.

Environmental Standards

In preparing an EA for an energy project, sponsors must demonstrate that the environmental protection plan conforms with established standards. These standards refer to the limits and boundaries of environmental risks. For example, the maximum emission of sulfur dioxide from a power plant should stay below a standard, to ensure that the ambient air quality does not fall below an acceptable level. The most frequently used standards relate to air and water quality. Many countries have two sets of air quality standards, one referring to ambient air and the other to allowable emissions. In the same way, standards of water quality refer to ambient water quality and allowable effluent levels. The ambient standards for air and water are set at levels below which no significant damage to the environment is likely. Emission levels for air and effluent levels for water represent the maximum allowable discharge from a point. These standards may be set at different levels, depending on prevailing ambient conditions, location of the plant, and so on.

For most developing countries, environmental standards are of recent origin. Indeed, many countries do not have environmental standards. However, worldwide attention to environmental concerns has forced most financiers to require that a proposed energy project meet acceptable environmental standards. As a result, most developing countries are devising standards based on the experiences of industrialized countries and international agencies.

Environmental standards of industrialized countries vary substantially in coverage and application. In some countries, standards depend on the size, type, and location of the plant;

other countries have different standards for existing versus new plants and for various seasons of the year. Thus, no unique set of standards exists for developing countries. In addition, environmental standards of industrialized countries are often viewed as too stringent for use in developing countries.

Although there are no generally accepted international environmental standards, sets of standards are issued by the EU, the World Health Organization, and other multilateral organizations. Financiers reviewing energy projects also use standards. The latter standards are most relevant to project preparation. Indeed, most EAs refer to these standards to establish environmental viability of a proposed project.

Environmental standards used by various financiers are often similar to the standards used by the World Bank Group. The World Bank does point out that its standards should be viewed as guidelines only and judged in the context of plant location and existing environmental conditions. The standards show the maximum levels of hazardous air emissions and liquid effluents for energy projects. Because of the basic differences in technologies and processes, the standards are different for oil and gas development from those for power generation projects. Some standards, however—for example, those for workplace air quality and noise level and for handling, storage, and safety of materials—are almost the same for oil, gas, and power projects.

Oil and gas development

The environmental standards and guidelines for oil and gas development include recommendations for changing certain processes and specify limits for air emissions and liquid effluents. (See table 11–1.)

Table 11–1. Air and Liquid Emission Parameters for Oil and Gas Projects

The air emission limits are as follows:

Parameter	Maximum value
Volatile organic compounds (VOCs; including benzene)	20 milligrams (mg) per normal cubic meter (Nm^3) (mg/Nm^3)
Hydrogen sulfide (H_2S)	5 mg/Nm^3
Nitrogen oxides (NO_x)	
Gas-fired	240 mg/Nm^3 (or 65 nanograms per joule [ng/J]) (for gaseous fuels)
Oil-fired	360 mg/Nm^3 (or 100 ng/J) (for liquid fuels)
Odor	Not offensive at the receptor end

The liquid effluent standards are as follows:

Parameter	Maximum value
pH	6 to 9
BOD_5	50 mg/l
COD	250 mg/l
Total suspended solids	50 mg/l
Oil and grease	10 mg/l
Phenol	0.5 mg/l
Total residual chlorine	1 mg/l
Cyanide (total)	0.2 mg/l
Arsenic	0.1 mg/l
Cadmium	0.1 mg/l
Chromium (hexavalent)	0.1 mg/l
Lead	0.1 mg/l
Mercury	0.001 mg/l
Total metals	10 mg/l
Temperature increase	less than 3°C[a]

[a] *The effluent should result in a temperature increase of no more than 3 degrees Celsius at the edge of the zone where initial mixing and dilution take place. Where the zone is not defined, use 100 meters from the point of discharge.*

Summary 11–4. Environmental Standards

- In preparing environmental assessments (EAs) for energy projects, sponsors need to demonstrate that environmental impacts do not exceed acceptable standards.
- Most developing countries prepare standards based on the practices of industrialized countries and the guidelines used by international agencies.
- Although no generally accepted international environmental standards exist, several sets of standards—issued by the European Union, the World Health Organization, and multilateral development institutions—are available.
- Environmental standards used by financiers are often similar to standards issued by the World Bank Group. The standards are provided separately for oil and gas projects and for power projects. They describe (1) recommended practices and processes, (2) maximum levels of hazardous air emissions and liquid effluents, and (3) general requirements for workplace air quality, noise level, among other concerns.

The recommended process changes are as follows:

- Maximize use of freshwater gel-based mud systems
- Eliminate use of invert muds
- Recycle drilling-mud decant water

- Prevent degradation of sweet wells by sulfate-reducing bacteria by using biocides
- Select less toxic biocides, corrosion inhibitors, and other chemicals
- Minimize gas flaring
- Store crude oil in tanks equipped with secondary seals and vapor-recovery systems
- Remove hydrogen sulfide and mercaptans from gases before flaring
- Use knockout drums on flares to prevent emissions of condensates
- Regenerate spent amines and spent solvents or send them off-site for recovery
- Use low-nitrogen-oxide (low-NO_x) burners in process heaters
- Provide spill prevention and control measures (incorporation of bunds and hard surfacing for areas around storage tanks, use of pressure-relief valves, and installation of high-level alarms)
- Recover oil from process wastewaters
- Segregate stormwater from process water

Other requirements are as follows:

- Hydrogen sulfide monitors must be installed wherever the gas may accumulate. The monitors should be set to activate warning signals whenever detected H_2S concentrations exceed 7 mg/m^3 (5 ppm).
- An assessment of oil spill risks must be conducted, and an oil spill contingency plan must be developed and implemented.

Oil and gas pipelines

Environmental guidelines specific to oil pipelines include

- Positive pipe corrosion control measures
- A program of periodic inspection and maintenance
- Pressure sensors connected to alarms and automatic pump shutdown systems
- A metering system that provides continuous input/output comparison for leak detection
- An engineering design that provides adequate protection from likely external physical forces
- Accurate and complete records of all inspections, leak incidents, unusual events, and safety measures taken
- Minimization of disturbance to natural vegetation, soils, hydrological regimes, and topography
- Positive measures to control population influx to remote areas because of the increased access created by the pipeline ROW and to prevent associated secondary impacts (e.g., encroachment on the lands or preserves of traditional indigenous populations and uncontrolled exploitation of natural resources)

Refineries

Emissions. Control of air emissions normally includes the capturing and recycling or burning of emissions from vents, product transfer points, storage tanks, and other handling equipment. Boilers, heaters, and other combustion devices;

cokers; and catalytic units may require controls for particulate matter (PM). Catalytic cracking units should be provided with particulate-removal devices. (See table 11–2.)

Table 11–2. Emission parameters for refineries

Parameter	Maximum Value*
PM	50 mg/Nm^3
NO_x	360 mg/Nm^3
SO_x	500 mg/Nm^3
H_2S	15 mg/Nm^3

*mg/Nm^3 = milligrams per normal cubic meters.

Effluents. Refinery wastewater often requires a combination of treatment methods to remove oil and other contaminants before discharge. Separation of different streams (e.g., stormwater) is essential to minimize treatment requirements. Oil is recovered from slops through separation techniques such as centrifuging. A typical system may include gravity separation, dissolved-air flotation, biodegradation, and clarification. A final polishing step using filtration or treatment with activated carbon or chemicals may also be required.

Power projects

Emissions. The following guidelines apply to new fossil fuel–fired thermal power plants or units of 50 megawatt electric (MWe) or larger and have been set at levels that can be achieved by adopting a variety of low-cost options or technologies, including the use of clean fuels. For example, dust controls capable of 98%–99% removal efficiency, such as electrostatic precipitators (ESPs) or baghouses, should always be installed. Similarly, the use of low-NO_x burners, usually with other combustion modifications, should be standard practice. The range of options for the control of SO_x is greater because of large differences in the sulfur content (% S) of different fuels and in control costs. In general, for low-sulfur, high-calorific fuels (<1% S), specific controls may not be required, whereas coal cleaning (when applicable), sorbent injection, or fluidized-bed combustion may be adequate for medium-sulfur fuels (1%–3% S). Flue-gas desulfurization (FGD) or other clean coal technologies should be considered for high-sulfur fuels (>3% S).

Particulates. For a coal-fired plant or unit, the recommended removal efficiencies from exhaust gases are 99% for all PM and 98% for PM_{10} (particulates of 10 microns or less in size). These removal efficiencies are to be achieved at least 95% of the time that the plant or unit is operating. For all plants or units, emissions should not exceed 50 milligrams per normal cubic meter (mg/Nm^3) for particulates under full-load conditions.

Nitrogen oxides. For a coal-fired plant or unit, a reduction in NO_x emissions of 40% (relative to the case in which no NO_x controls are installed) is recommended for 95% of the time that the plant or unit is operating. This should correspond to an emissions level of 650 mg/Nm^3 (230 nanograms per joule [ng/J]). For oil- and gas-fired plants or units, the recommended reduction rates are 25% and 5%, respectively, corresponding to emissions levels of 360 mg/Nm^3 (100 ng/J) and 240 mg/Nm^3 (65 ng/J).

Sulfur oxides. It is recommended that total SO_x emissions from the power plant or unit should be less than 0.20 tons per day (t/d) MWe of capacity for the first 1,000 MWe plus 0.01 t/d MWe for the incremental over 1,000 MWe. The concentration of SO_x in flue gases should not exceed 2,000 mg/Nm^3.

Effluents. The standards for effluents given in table 11–3 apply to all types of thermal power plants. Other environmental requirements for power projects include the following:

- Formulations containing chromates should be avoided in water-treatment processes.
- Transformers or equipment containing polychlorinated biphenyls (PCBs) or PCB-contaminated oil should not be installed, and existing equipment using these chemicals should be phased out and disposed of in a manner consistent with the requirements of the host country.
- Processes, equipment, and central cooling systems involving the use or potential release to the environment of chlorofluorocarbons (CFCs) and halons should not be installed, and these chemicals should be phased out of use in existing systems and processes and disposed of in a manner consistent with the requirements of the host country.
- Storage and liquid impoundment areas for fuels, raw and in-process materials, solvents, wastes, and finished products should be designed with secondary containment (e.g., dikes and berms) to prevent spills and contamination of soil, groundwater, and surface waters.

TABLE 11–3. EFFLUENT PARAMETERS FOR ALL TYPES OF THERMAL POWER PLANTS

Parameter	Maximum value
pH	6 to 9
Total suspended solids	50 mg/l
Oil and	0.5 mg/l
Chromium (hexavalent)	0.1 mg/l
Copper	0.5 mg/l
Iron	1.0 mg/l
Nickel	0.5 mg/l
Zinc	0.5 mg/l
Temperature increase	less than or equal to 3° C[a]

[a] *The effluent should result in a temperature increase of no more than 3 degrees Celsius at the edge of the zone where initial mixing and dilution take place. Where the zone is not defined, use 100 meters from the point of discharge.*

Health and safety measures recommended to prevent electrocution include the following:

- Strict procedures must be followed for de-energizing and checking electrical equipment before maintenance work.
- Strict safety procedures must be implemented, including constant supervision during maintenance work on energized equipment.
- Personnel must be trained on revival techniques in case of electrocution.

General requirements for oil, gas, and power projects

Workplace air quality. The standards for workplace air quality are as follows:

- Periodic monitoring of workplace air quality should be conducted for air contaminants relevant to employees' tasks and the plant's operations.
- Ventilation, air contaminant control equipment, protective respiratory equipment, and air quality–monitoring equipment should be well maintained.
- Protective respiratory equipment must be used by employees when the exposure levels for welding fumes, solvents, and other materials present in the workplace exceed local or internationally accepted standards or the following threshold limit values (TLVs):

  – CO, 29 mg/m^3

  – H_2S, 14 mg/m^3

  – N_2O, 5 mg/m^3

  – S_2O, 5 mg/m^3

- Monitors should be installed that activate an audible alarm when toxic gas concentrations exceed half the TLV.

Workplace noise. The requirements for limiting workplace noise are as follows:

- Feasible administrative and engineering controls, including sound-insulated equipment and control rooms, should be employed to reduce the average noise level in normal work areas.

- Plant equipment should be well maintained to minimize noise levels.
- Personnel must use hearing protection when exposed to noise levels above 85 dB.

Work in confined space. The requirements for work in confined spaces are as follows:

- All confined spaces (e.g., tanks, sumps, vessels, sewers, and excavations) must be tested for the presence of toxic, flammable, and explosive gases or vapor and for the lack of oxygen before individuals enter or occupy these spaces.
- Adequate ventilation must be provided to these spaces before workers enter them and during the time they occupy these spaces.
- Personnel must use air-supplied respirators when working in confined spaces that may become contaminated or deficient in oxygen during the period of occupancy.
- Observers and assistants must be stationed outside of confined spaces to provide emergency assistance, if necessary, to personnel working inside these areas.

Hazardous material handling and storage. The guidelines for handling and storing hazardous materials are as follows:

- All hazardous (reactive, flammable, radioactive, corrosive, and toxic) materials must be stored in clearly labeled containers or vessels.
- Storage and handling of hazardous materials must be in accordance with local regulations and appropriate to their hazard characteristics.

- Fire prevention systems and secondary containment should be provided for storage facilities, where necessary or required by regulation, to prevent fires or the release of hazardous materials to the environment.

General health. The guidelines for general health are as follows:

- Sanitary facilities should be well equipped with supplies (e.g., protective creams), and employees—particularly those exposed to dust, chemicals, or pathogens—should be encouraged to wash frequently.
- Ventilation systems should be provided to control work-area temperature and humidity.
- Personnel required to work in areas of high temperature or high humidity should be allowed to take frequent breaks away from these areas.
- Preemployment and periodic medical examinations should be conducted for all personnel, and specific surveillance programs should be instituted for personnel potentially exposed to toxic or radioactive substances.

General safety. The general safety precautions and measures are as follows:

- Shield guards or guard railings should be installed at all belts, pulleys, gears, and other moving parts.
- Elevated platforms and walkways, stairways, and ramps should be equipped with handrails, toeboards, and nonslip surfaces.
- Electrical equipment should be grounded and well insulated and in conformance with applicable codes.

- Personnel should use special footwear, masks, and clothing for work involving hazardous materials.
- For work near molten or high-temperature materials, employees should be provided with nonslip footwear, gloves, safety glasses, helmets, face protection, leggings, and other necessary protective equipment.
- Eye protection should be worn by personnel where there is a risk of flying chips or sparks or where intense light is generated.
- Personnel should wear protective clothing and goggles where corrosive materials are stored or processed.
- Emergency eyewash and shower stations should be installed in areas containing corrosive materials.
- A safety program should be established for construction and maintenance work.
- A fire prevention and safety program should be implemented and should include regular fire drills.

Training. The guidelines for training are as follows:

- Employees should be trained to recognize the hazards presented by all potentially harmful materials relevant to each employee's task and work area, as well as the precautions and procedures for the safe storage, handling, and use of those materials.
- Training should incorporate information from the material safety data sheets (MSDSs) for potentially harmful materials.

- Personnel should be trained in environmental, health, and safety matters, including accident prevention, safe lifting practices, the use of MSDSs, safe chemical-handling practices, and proper control and maintenance of equipment and facilities.
- Training should also include instructions on responding to an emergency, including the location and proper use of emergency equipment, the use of personal protective equipment, procedures for raising an alarm and notifying emergency response teams, and proper response actions for all foreseeable emergency situations.

Appendix A:

The Coordinates for Finance and Guarantee Agencies

Multilateral Cofinanciers

The Arab Bank for Economic Development in Africa (BADEA)
Sayed Abdel Rahman El-Mahdi Avenue
PO Box 2640
Khartoum, Sudan
Tel.: +249 1 83773646
Fax: +249 1 83770600
http://www.badea.org

The Arab Fund for Economic and Social Development
PO Box 21923
Safat 13080 Kuwait City
State of Kuwait
Tel.: +965 495 9000
Fax: +965 48 15750
http://www.arabfund.org

The African Development Bank (AfDB)
Rue Joseph Anoma
01 BP 1387 Abidjan 01
Côte d'Ivoire
Tel.: +225 20 20 44 44
Fax: +225 20 20 49 59
http://www.afdb.org

The African Development Fund (AfDF)
Rue Joseph Anoma
01 BP 1387 Abidjan 01
Côte d'Ivoire
Tel.: +225 20 20 44 44
Fax: +225 20 20 49 59
http://www.afdb.org

The Asian Development Bank (ADB)
6 ADB Avenue
Mandaluyong City
1550 Metro Manila
Philippines
Tel.: +632 632 4444
Fax: + 632 636 2444
http://www.adb.org

The European Investment Bank (EIB)
100 boulevard Konrad Adenauer
L-2950 Luxembourg
Tel.: +352 43 79 1
Fax: +352 43 77 04
http://www.eib.europa.eu

The European Bank for Reconstruction and Development (EBRD)
One Exchange Square
London EC2A 2JN
United Kingdom
Tel.: +44 20 7338 6000
Fax: +44 20 7338 6100

The Inter-American Development Bank (IDB)
1350 New York Avenue NW
Washington, DC 20577
USA
Tel.: +1 202 623 3900
Fax: +1 202 623 3815
http://www.iadb.org

The International Finance Corporation (IFC)
2121 Pennsylvania Avenue NW
Washington, DC 20433
USA
Tel.: +1 202 473 3800
Fax: +1 202 974 4384
http://www.ifc.org

The Islamic Development Bank (IsDB)
PO Box 5925
Jeddah 21432
Kingdom of Saudi Arabia
Tel.: +9662 636 1400
Fax: +9662 636 6871
http://www.isdb.org

The Nigeria Trust Fund (NTF)
Rue Joseph Anoma
01 BP 1387 Abidjan 01
Côte d'Ivoire
Tel.: +225 20 20 44 44
Fax: +225 20 20 49 59
http://www.afdb.org

The Nordic Development Fund (NDF)
Fabianinkatu 34
PO Box 185
FIN-00171 Helsinki
Finland
Tel.: +358 9 180 0451
Fax: +358 9 622 1491
http://ww.ndf.fi

The Nordic Investment Bank (NIB)
Fabianinkatu 34
PO Box 249
FI-00171 Helsinki
Finland
Tel.: +358 9 18 001
Fax: +358 9 1800 210
http://www.nib.int

The OPEC Fund for International Development
PO Box 995
A-1011 Vienna
Austria
Tel.: +43 1 515 64 0
Fax: +43 1 513 92 38
http://www.opecfund.org

The World Bank (IBRD)
1818 H Street NW
Washington, DC 20433
USA
Tel.: +1 202 473 1000
Fax: +1 202 477 6391
www.worldbank.org

Bilateral Cofinanciers

Australia

Australian Agency for International Development (AusAid)
GPO Box 887
Canberra ACT 2601
Australia
Tel.: +61 2 6206 4000
Fax: +61 2 6206 4880
http://www.ausaid.gov.au

Export Finance and Insurance Corporation (EFIC)
Level 10 Export House
22 Pitt Street
Sydney NSW 2000
Australia
Tel.: +61 2 9201 2111
Fax: +61 2 9251 3851
http://www.efic.gov.au

Austria

Austrian Development Agency (ADA)
Zelinkagasse 2
A-1010 Vienna
Austria
Tel.: +43 1 90 3 990
Fax: +43 1 90 3 99 290
http://www.ada.gv.at

Austrian Export Credit Agency (OEKB)
Am Hof 4
Strauchgasse 3
1011 Vienna
Austria
Tel.: +43 1 531 27 2859
Fax: +43 1 531 27 5698
http://www.oekb.at

Belgium

Belgian Corporation for International Investment (BMI-SBI)
Avenue de Tervurenlaan 168/b.9
1150 Brussels
Belgium
Tel.: +32 2 776 01 00
Fax: +32 2 770 66 38
http://www.bmi-sbi.be

Belgian Export Credit Agency (ONDD)
Rue Montoyerstraat 3
1000 Brussels
Belgium
Tel.: +32 2 788 88 00
Fax: +32 2 788 88 10
http://www.ondd.be

Belgian Technical Cooperation (BTC)
Rue Haute 147
1000 Brussels
Belgium
Tel.: +32 2 505 37 00
Fax: +32 2 502 98 62
http://www.btcctb.org

Directorate-General for Development Cooperation (DGDC)
Rue des Petits Carmes 15
B-1000 Brussels
Belgium
Tel.: +32 2 501 81 11
http://www.dgcd.be

Canada

Canadian International Development Agency (CIDA)
200 Promenade du Portage
Gatineau, QC K1A 0G4
Canada
Tel.: +1 819 997 5006
Fax: +1 819 953 6088
http://www.acdi-cida.gc.ca

Export Development Canada (EDC)
151 O'Connor
Ottawa, ON K1A 1K3
Canada
Tel.: +1 613 598 2500
Fax: +1 613 237 2690
http://www.edc.ca

Czech Republic

Czech Export Bank
Vodičova 34
PO Box 870
111 21 Praha 1
Czech Republic
Tel.: +420 222 843 111
Fax: +420 224 225 162
http://www.ceb.cz

Ministry of Foreign Affairs
Loretánské náměstí 5
118 00, Praha 1
Czech Republic
Tel.: +224 181 111
http://www.czechembassy.org

Denmark

Danish International Development Agency (DANIDA)
Royal Danish Ministry of Foreign Affairs
Asiatisk Plads 2
1448 Copenhagen
Denmark
Tel.: +45 33 92 00 00
Fax: +45 32 54 05 33
http://www.um.dk

Eksport Kredit Fonden (EKF)
Dahlerups Pakhus, Langelinie Allé 17
2100 Copenhagen
Denmark
Tel.: +45 35 46 61 00
http://www.ekf.dk/

Finland

Finnvera
Eteläesplanadi 8
00130 Helsinki
Finland
Tel.: +358 20 460 11
Fax: +358 204 60 7220
http://www.finnvera.fi

Ministry for Foreign Affairs
Merikasarmi, PO Box 176
00161 Helsinki, Finland
Tel.: +358 9 160 05
Fax: +358 9 629 840
http://formin.finland.fi

France

COFACE
92065 Paris La Défense Cedex
France
Tel.: +33 1 47 52 41 00
http://www.coface.com

French Development Agency (AFD)
5 rue Roland Barthes
75598 Paris Cedex 12
France
Tel.: +33 1 53 44 31 31
Fax: +33 1 44 87 99 39
http://www.afd.fr

Germany

Agency for Technical Co-operation (GTZ)
Postfach 5180
65726 Eschborn
Germany
Tel.: +49 6196 790
Fax: +49 6196 79 1115
http://www.gtz.de

Federal Ministry for Economic Cooperation and Development (BMZ)
Adenauerallee 139–141
53113 Bonn
Germany
Tel.: +49 18 88 535 0
Fax: +49 18 88 535 3500
http://www.bmz.de

KfW Development Bank
KfW Bankengruppe
Charlottenstrasse 33/33a
10117 Berlin
Germany
Tel.: +49 69 74313363
http://www.kfw-entwicklungsbank.de

Greece

Export Credit Insurance Organization (ECIO)
57 Panepistimiou Street
10564 Athens
Greece
Tel.: +30 210 3310017 20
Fax: +30 210 3244074
http://www.ecio.gr

YDAS (Hellenic Aid) under the MOFA
5 Vassilissis Sophias Avenue
106 71 Athens
Greece
Tel.: +30 210 3681000
http://www.mfa.gr

Hungary

Hungary Export Import Bank
H-1065 Budapest, Nagymezőutca 46.-48.
Hungary
Tel.: +36 1 374 9100
Fax: +36 1 269 4476
http://www.eximbank.hu

Ministry of Foreign Affairs
1027 Budapest, Bem rakpart 47.
Hungary
Tel.: +36 1 458 1000
Fax: +36 1 212 5918
http://www.mfa.gov.hu

Iceland

Icelandic International Development Agency (ICEIDA)
Thverholt 14
125 Reykjavík
Iceland
Tel.: +354 545 8980
Fax: +354 545 8985
http://www.iceida.is

Ireland

The Insurance Corporation of Ireland (ICI)
Burlington Road
Dublin 4
Ireland
Tel.: +353 1 702 3000
Fax: +353 1 660 9220

Irish Aid, Department of Foreign Affairs
Bishops Square, Redmond Hill
Dublin 2
Ireland
Tel.: +353 1 408 2000
Fax: +353 1 408 2880
http://www.irishaid.gov.ie

Italy

Ministry of Foreign Affairs (DGCS)
Piazzale della Farnesina
1-00194 Rome
Italy
Tel.: +39 06 36911
http://www.esteri.it

SACE S.p.A. (Servizi Assicurativi del Commercio Estero)
Piazza Poli
37/42, Sede Centrale, Rome
Italy
Tel.: +39 06 67361
Fax: +39 06 6736225
http://www.sace.it

Japan

Japan Bank for International Cooperation (JBIC)
4-1 Ohtemachi 1-chome, Chiyoda-ku
Tokyo 100-8144
Japan
Tel.: +81 03 5218 3101
Fax: +81 03 5218 3955
http://www.jbic.go.jp

Japan International Cooperation Agency (JICA)
6th–13th Floors, Shinjuku Maynds Tower
2-1-1 Yoyogi, Shibuya-ku
Tokyo 151-8558
Japan
Tel.: +81 03 5352 5311
http://www.jica.go.jp

Nippon Export and Investment Insurance (NEXI)
Chiyoda First Building, East Wing, 3rd Floor
3-8-1 Nishikanda, Chiyoda-ku,
Tokyo 101-8359
Japan
Tel.: +81 03 3512 7650
Fax: +81 03 3512 7660
http://nexi.go.jp

Kuwait

Kuwait Fund for Arab Economic Development
Mirqab
Mubarak Al-Kabeer Street
PO Box 2921
Safat 13030 Kuwait City
State of Kuwait
Tel.: +965 2999000
Fax: +965 2999090
http://www.kuwait-fund.org

Luxembourg

Lux-Development (Luxembourg Agency for Development Cooperation)
S.A. B.P. 2273 L-1022
Luxembourg
Tel.: +352 29 58 58 1
Fax: +352 29 58 58 200
http://www.lux-development.lu

l'Office du Ducroire
7 rue Alcide de Gasperi L-2981
Luxembourg
Tel.: +352 42 39 39 320
http://www.ducroire.lu

Netherlands

Ministry of Foreign Affairs
Bezuidenhoutseweg 67
The Hague
Netherlands
Tel.: +31 70 3486486
Fax: +31 70 3484848
http://www.minbuza.nl

New Zealand

New Zealand Export Credit Office (NZECO)
The Treasury
1 The Terrace
PO Box 3724
Wellington
New Zealand
Tel.: +64 4 471 5021
Fax: +64 4 499 0437
http://www.nzeco.govt.nz

New Zealand International Aid and Development Agency (NZAID)
195 Lambton Quay, Private Bag 18-901
Wellington
New Zealand
Tel.: +64 4 439 8200
Fax: +64 4 439 8515
http://www.nzaid.govt.nz

Norway

Eksportfinans
Dronning Maudsgate 15
PO Box 1601
Vika. 0119 Oslo
Norway
Tel.: +47 22 01 22 01
Fax: +47 22 01 22 02
http://www.eksportfinans.no

Ministry of Foreign Affairs, International Development Cooperation
7. juni-plassen/Victoria Terrasse
PO Box 8114
Dep. N-0032 Oslo
Norway
Tel.: +47 22 24 50 55
Fax: +47 22 24 95 20
http://odin.dep.no

Norwegian Agency for Development Cooperation (Norad)
Ruseløkkveien 26
0251 Oslo
Norway
Tel.: +47 22 24 20 30
Fax: +47 22 24 20 31
http://www.norad.no

Norwegian Guarantee Institute for Export Credits (GIEK)
Dronning Maudsgate 15
PO Box 1763
Vika. N-0122 Oslo
Norway
Tel.: +47 22 87 62 00
Fax: +47 22 83 24 45
http://www.giek.no

POLAND

PolishAid
Ministry of Foreign Affairs of the Republic of Poland
Development Co-operation Department
Al. J. Ch. Szucha 23
00-580 Warsaw
Poland
Tel.: +48 22 523 8073
Fax: +48 22 523 8074
http://www.polishaid.gov.pl

Export Credit Insurance Corporation
ul. Sienna 39, 00-121 Warszawa
Tel.: +48 22 356 83 00
Fax: +48 22 313 01 19
http://www.kuke.com.pl

Portugal

Companhia de Seguro de Creditos, S.A. (COSEC)
Avenida da República, nº 58
1069-057 Lisboa
Portugal
http://www.cosec.pt

Portuguese Institute for Development Support (IPAD)
Avenida da Liberdade, nº 192-2º
1250-147 Lisboa
Portugal
Tel.: +351 213176700
Fax: +351 213147897
http://www.ipad.mne.gov.pt

Saudi Arabia

Saudi Fund for Development (SFD)
PO Box 50483
Riyadh
Kingdom of Saudi Arabia 11523
Tel.: +00966 1 4640292
Fax: +00966 1 4647450
http://www.sfd.gov.sa/

Slovak Republic

Agency for International Development Cooperation—Slovak Aid
Hlboká cesta 2
833 36 Bratislava 37
Slovak Republic
Tel.: +421 2 5978 2741
Fax: +421 2 5978 3579
http://www.slovakaid.mfa.sk

Eximbanka SR
Grösslingová 1
813 50 Bratislava
Slovak Republic
Tel.: +421 2 59 398 111
http://www.eximbanka.sk

SOUTH KOREA

Economic Development Cooperation Fund (EDCF)
Eximbank, 16-1, Yeouido-dong, Yeongdeungpo-gu
Seoul 150-996
South Korea
Tel.: +82 2 3779 6114
Fax: +82 2 784 1030
http://www.koreaexim.go.kr

Korea Export Insurance Corporation (KEIC)
2nd–16th Floors, Seoul Central Building
136 Seorin-dong, Jongro-ku
Seoul 110-729
South Korea
Tel.: +82 2 399 6800
Fax: +82 2 399 6598
http://www.keic.or.kr

Korea International Cooperation Agency (KOICA)
110-782 #128, Yunkun-dong, Jongno-gu,
Seoul
South Korea
Tel.: +82 2 740 5114
Fax: +82 2 740 1092
http://www.koica.go.kr

Spain

Compañía Española de Crédito a la Exportación (CESCE)
C/ Velázquez, 74
28001 Madrid
Spain
Tel.: +91 902 11 10 10
http://www.cesce.es

Instituto Español de Comercio Exterior (ICEX)
Pº de la Castellana 14-16
28046 Madrid
Spain
Tel.: +91 902 349 000
http://www.icex.es

Secretaría de Estado de Cooperación Internacional (SECI)
MAEC, Plaza de la Provincia
1-28012 Madrid
Spain
Tel.: +91 583 81 00
http://www.mae.es

Sweden

Export Credit Guarantee Board (EKN)
Kungsgatan 36
Box 3064
SE-103 61 Stockholm
Sweden
Tel.: +46 8 788 00 00
Fax: +46 8 411 81 49
http://www.ekn.seSwedish Export Credit Corporation

AB Svensk Exportkredit
Västra Trädgårdsgatan 11 B,
SE-103 27 Stockholm
Sweden
Tel.: +46 8 613 83 00
Fax: +46 8 20 38 94
http://www.sek.se

Swedish International Development Cooperation Agency (SIDA)
Valhallavägen 199
105 25 Stockholm
Sweden
Tel.: +46 8 698 50 00
Fax: +46 8 20 88 64
http://www.sida.se

Switzerland

State Secretariat for Economic Affairs SECO Economic Development Cooperation
Effingerstrasse 1
3003 Bern
Switzerland
Tel.: +41 31 324 09 15
Fax: +41 31 324 14 90
http://www.seco-cooperation.ch

Swiss Agency for Development and Cooperation (SDC)
Head Office
Freiburgstrasse 130
3003 Bern
Switzerland
Tel.: +41 31 322 34 75
Fax: +41 31 324 16 94
http://www.deza.ch

Swiss Export Risk Insurance (SERV)
Kirchenweg 8
PO Box CH-8032
Zurich
Switzerland
Tel.: +41 44 384 4777
Fax: +41 44 384 4787
http://www.serv-ch.com

Turkey

Export Credit Bank of Turkey (Turk Eximbank)
Müdafaa Cad No. 20
06100 Ankara
Tel.: +90 312 417 13 00
Fax: +90 312 425 78 96
http://www.eximbank.gov.tr

Ministry of Foreign Affairs
Balgat 06100 Ankara
Turkey
Tel.: +90 312 292 10 00
http://www.mfa.gov.tr

United Arab Emirates

Abu Dhabi Fund for Arab Economic Development
PO Box 814
Saees Bin Ghobash Building
Abu Dhabi
United Arab Emirates
Tel.: +97 12 725 800
Fax: +97 12 728 890

United Kingdom

Department for International Development (DFID)
1 Palace Street
London SW1E 5HE
United Kingdom
Tel.: +44 1355 84 3132
Fax: +44 1355 84 3632
http://www.dfid.gov.uk

Export Credits Guarantee Department (EGGD)
2 Exchange Tower, Harbour Exchange Square
PO Box 2200
London E14 9GS
United Kingdom
Tel.: +44 20 7512 7000
Fax: +44 20 7512 7649
http://www.ecgd.gov.uk

UK Trade and Investment (UKTI)
Tay House
300 Bath Street
Glasgow G2 4DX
United Kingdom
Tel.: +44 20 7215 8000
https://www.uktradeinvest.gov.uk

United States

Export-Import Bank of the United States
811 Vermont Avenue NW
Washington, DC 20571
USA
Tel.: +1 202 565 3946
http://www.exim.gov

Millennium Challendge Corporation (MCC)
875 15th Street NW
Washington, DC 20005-2221
USA
Tel.: +1 202 521 3600
http://www.mcc.gov

Overseas Private Investment Corporation
1100 New York Avenue NW
Washington, DC 20527
USA
Tel.: +1 202 336 8400
http://www.opic.gov

Private Export Funding Corporation (PEFCO)
280 Park Avenue
New York, NY 10017
USA
Tel.: +1 212 916 0300
Fax: +1 212 286 0304
http://www.pefco.com

U.S. Agency for International Development (USAID)
Ronald Reagan Building

Washington, DC 20523-1000
USA
Tel.: +1 202 712 4810
Fax: +1 202 216 3524
http://www.usaid.gov

U. S. Trade and Development Agency (TDA)
1000 Wilson Boulevard, Suite 1600
Arlington, VA 22209
USA
Tel.: +1 703 875 4357
Fax: +1 703 875 4009
http://www.tda.gov

Prominent Commercial Financiers

ABN AMRO Bank N.V.
Corporate and Institutional Banking
Head Office
Gustav Mahlerlaan 10
1082 PP Amsterdam
Netherlands
Tel.: +31 20 628 98 98
http://www.abnamro.com

Allianz Group
Königinstraße 28
D-80802 Munich
Germany
Tel.: +49 89 3800 0
Fax: +49 89 3800 3425
http://www.allianz.com

Bank of America
Corporate Center
100 North Tryon Street
Charlotte, NC 28255
USA
Tel.: +1 800 432 0000
http://www.bankofamerica.com

BNP Paribas
3 rue d'Antin
75078 Paris Cedex 02
France
Tel.: +33 1 40 14 63 58
Fax : +33 1 42 98 21 22
http://www.bnpparibas.com

Calyon
9 quai du Président Paul Doumer
92920 Paris La Défense Cedex
France
Tel.: +33 1 41 89 00 00
http://www.calyon.com

CIBC
Head Office
5650 Yonge Street
Toronto, ON M2M 4G3
Canada
Tel.: +1 416 980 2211
Fax: +1 416 363 5347
www.cibc.com

Citigroup
399 Park Avenue
New York, NY 10043
USA
Tel.: +1 800 285 3000
www.citigroup.com

Credit Suisse Group
Paradeplatz 8
8070 Zurich
Switzerland
Tel.: +41 44 212 16 16
Fax: +41 44 333 25 87
http://www.credit-suisse.com

Deutsche Bank AG
Taunusanlage 12
60325 Frankfurt am Main
Germany
Tel.: +49 69 910 00
Fax: +49 69 910 34225
http://www.deutsche-bank.de

Fortis SA/NV
Rue Royale 20
1000 Brussels
Belgium
Tel.: +32 2 565 11 11
Fax: +32 2 565 42 22
http://www.fortis.com

HSBC Group
Head Office, HSBC Holdings PLC
8 Canada Square
London E14 5HQ
United Kingdom
Tel.: +44 20 7991 8888
http://www.hsbc.com

ING
Amstelveenseweg 500
1081 KL Amsterdam
Netherlands
Tel.: +31 20 541 5411
Fax: +31 20 541 5497
http://www.ing.com

Mitsubishi UFJ
7-1 Marunouchi 2-Chome, Chiyoda-ku
Tokyo
Japan
http://www.mufg.jp

Mizuho Bank
1-1-5 Uchisaiwaicho, Chiyoda-ku
Tokyo 100-0011
Japan
Tel.: +81 3 3596 1111
http://www.mizuhobank.co.jp

Rabobank
PO Box 17100 (UC-R315)
3500 HG Utrecht
Netherlands
Tel.: +31 30 2165777
Fax: +31 30 2161863
http://www.rabobank.com

Santander Group
Ciudad Grupo Santander
Avenida de Cantabria s/n, Edificio Pereda, 1ª planta
28014 Madrid
Spain
Tel.: +34 91 259 65 14
http://www.santander.com

Societe General
Tour Société Générale
17 cours Valmy
92987 Paris La Défense Cedex
France
Tel.: +33 1 42 14 20 00
http://www.sgcib.com

Sumimoto Mitsui Financial Group
1-2 Yurakucho 1-chome, Chiyoda-ku
Tokyo 100-0006
Japan
Tel.: +81 3 5512 3411
http://www.smfg.co.jp

UBS Investment Bank
Bahnhofstraße 45
PO Box CH-8098
Zurich
Switzerland
Tel.: +41 44 234 11 11
Fax: +41 44 239 91 11
http://www.ibb.ubs.com

Appendix B:

Guidelines for the Preparation of Environmental Assessments

Most financiers have published guidelines for the preparation of environmental assessments (EAs). These guidelines are similar to those issued by the World Bank Group. This appendix contains a summary of the World Bank guidelines. The details are published in a three-volume *Environmental Assessment Sourcebook* (1995), supplemented by notes available at the World Bank Web site.

Outline of an EA

An EA should include the following sections:

- Executive summary
- Introduction
- Legislative, regulatory, and policy considerations
- Description of the proposed project

- Description of the baseline environment
- Potential impacts of the proposed project
- Analysis of alternatives
- Environmental management and mitigation plan
- Monitoring plan
- Coordination with government agencies, affected communities, and nongovernmental organizations (NGOs)
- References

Executive summary

The executive summary must be a concise description of the proposed project, environmental setting, significant findings, and recommended actions to mitigate significant impacts and monitor environmental performance. The executive summary should consist of no more than 20 pages of text, with one figure or map showing the project's location.

Introduction

This section should contain the following items:

- Brief description of the general location and type of activity
- Identification of project sponsors
- Brief history of the project, if the project represents expansion, modernization, or restart of previous activity at the proposed site

Legislative, regulatory, and policy considerations

All applicable regulations and standards governing liquid effluents, air emissions, solid waste management, environmental quality, and health and safety must be identified and outlined with numerical standards specified in summary tables. These include national and local regulations and standards, as well as the guidelines and requirements of any financial institutions to which the sponsors apply for financing. The project's consistency with relevant World Bank safeguard policies must also be described when appropriate.

Description of the proposed project

The EA must provide a description of the project, preferably using maps (at appropriate scale), figures, and tables when appropriate and including the following information:

- Site(s), including any installations or investments of the project sponsor that are directly linked to the proposed project even though outside the main project site
- Provision of services from off-site (e.g., energy, water, or transportation)
- Process diagram, showing the steps in the process, material flow and balances, and sources of liquid effluents, air emissions, solid waste, and any other waste
- Locations of liquid effluent discharge points
- Information on the quality of the air emissions and liquid effluent after relevant treatment, together with a description of the treatment system or pollution control technology installed to achieve the specified waste characteristics
- Environmental alterations during construction (e.g., land grading, ROW clearance, and road building)

- Project employment
- Proposed organization of environmental management staff and associated training
- Projected occupational conditions related to worker health (e.g., noise levels and workplace air quality) and safety (e.g., hazardous areas, operations, or equipment)
- Proposed occupational health and safety programs, including training

DESCRIPTION OF THE BASELINE ENVIRONMENT

Detailed descriptions of baseline environmental conditions in and surrounding the operation must be provided. Graphic presentations of data (e.g., figures and tables) are preferred when practical. Environmental features to be described (focusing on those environmental features likely to be affected by the project) include

- Climate and air quality
- Landform (topography, geology, seismicity, and soils)
- Hydrology, water quality, and groundwater resources
- Ecology, flora, and fauna (including identification of unique or sensitive natural habitats and locally or internationally recognized endangered or threatened species)
- Existing land and water resource uses, including potentially affected areas adjacent to the project site and focusing particularly on areas and uses likely to be affected (e.g., receiving waters for liquid effluents)
- Archeological, historical, and culturally sensitive resources at the site

- Socioeconomic conditions, especially potentially affected or concerned communities and organizations

If the project is an expansion, modernization, or restart of existing activities, the EA must provide a summary of any significant, persistent environmental problems related to past or current operations of the industrial facilities. This portion of the EA should identify any areas in which current conditions do not comply with applicable national, local, or World Bank requirements and guidelines and should specify measures required in order to bring these areas into compliance.

Potential impacts of the proposed project

The EA must

- Provide a brief comparison of the impacts associated with alternative sites or processes, as appropriate, and key factors in decisions to select the proposed site and process
- Identify all significant environmental, socioeconomic, and human health and safety impacts associated with the project
- Distinguish between beneficial and adverse impacts, direct and indirect impacts, and immediate and long-term impacts on the environment
- Clearly identify significant impacts that are unavoidable or irreversible
- Wherever possible, describe impacts quantitatively, in terms of environmental costs and benefits, assigning economic values where feasible

- Characterize and explain significant information deficiencies and ensuing uncertainties associated with predictions of impact
- Describe the planned and implemented programs, or both, to inform concerned governmental organizations and NGOs, as well as local communities, and involve them in the selection and design of mitigation, compensation, and monitoring measures

In certain cases, cumulative impacts might need to be assessed. Such cases could arise when the project includes plans for significant future expansion or has the potential to interact with other proposed development activities such that the resulting impact on the environment is much greater than it would be if each occurred alone. As general guidance, preparers of the EA should determine, to the extent practical, whether other development projects are being considered for the project area that are likely to be implemented and that could interact substantively with the proposed project.

Analysis of alternatives

The EA should include an analysis of reasonable alternatives that meet the ultimate project objectives. These alternatives could include upgrading of existing facilities (rather than construction of a new plant), alternative routes and sites, alternative designs, and alternative methods of construction. They also should include the "no action" alternative—that is, not constructing the project, to demonstrate environmental conditions without the proposed project. The EA should then compare these alternatives in terms of capital and operating costs, skill requirements, political acceptability, public cooperation, and institutional requirements. To the extent possible, the EA should quantify the costs and benefits of various alternatives and describe the reasons for selecting the proposed project over other alternatives.

Environmental management and mitigation plan

The EA must include proposals for mitigation of any significant adverse impacts of the project and plans for ongoing management of the project to ensure that environmental impacts will be kept to a minimum throughout the lifetime of the project. This portion of the EA should

- Identify feasible and cost-effective measures to prevent significant adverse impacts or reduce them to acceptable levels
- Estimate capital and recurrent costs for environmental management activities
- Identify institutional and training requirements (i.e., development of capabilities needed to implement mitigation measures and monitoring programs)
- Propose work plans and schedules (to ensure that actions are taken in phase with appropriate project engineering activities)
- Specify compensatory measures where mitigative measures are not feasible or cost-effective

Monitoring plan

The EA must include a detailed plan for monitoring waste discharges, emissions, and environmental parameters in and around the project. The proposed monitoring scheme must allow determination of rates and concentrations of emissions and waste discharges from the project, occupational health and safety conditions, and the effectiveness of mitigating measures.

The monitoring plan should include an estimate of capital and operating costs and a description of other inputs (e.g., training and institutional strengthening) required for its implementation. The plan should also identify action levels for each parameter monitored, at which point action will be taken to reduce or mitigate associated impacts.

Specific elements to be included in this plan will vary according to the potential environmental and occupational health and safety concerns associated with each project. In many projects, elements of a monitoring program include

- Stack emissions and ambient air quality
- Effluents released to surface waters, including stormwater runoff and receiving-water quality
- Accident frequency and severity, temperature, noise, and air quality conditions at workstations
- Socioeconomic conditions

Coordination with Government Agencies, Affected Communities, and NGOs

As mentioned earlier, the sponsors should make an active effort during the preparation of the EA to consult in a culturally acceptable context with appropriate government agencies and obtain the views of local NGOs, affected communities, and other concerned groups. Records should be kept of meetings and other activities, communications, comments, and actions taken or project modifications made in response to public and community input. The EA should summarize steps taken to consult with and highlight key concerns of local interested parties and should include a brief explanation of how these concerns were addressed in the draft project design.

References

All information sources referenced in the text should be identified, with full citations in the case of published information. In the case of unpublished documents and personal communications, sufficient information should be provided to inform the reader about the sources of information. For example, references to personal communications should indicate the names and affiliations of communicating parties and the approximate date of the communication. References to unpublished reports should include the name and location (city and country) of the offices where such documents are located.

Template for Mitigation Measures and Monitoring of Thermal Power Projects

Preparation of an EA requires compilation of substantial amounts of information. It also requires intimate familiarity with international practices, particularly when preparing the mitigation and monitoring plans. The template given in tables B–1 and B–2 serves as a practical tool for the preparation of mitigation measures and monitoring for a thermal power project. It can be modified easily for other types of energy projects.

Table B–1. Thermal Power Project Template, Part I—Mitigation Measures

Project Activity	Potential Environmental Impacts	Proposed Mitigation Measure(s) (incl. legislation & regulations)	Institutional Responsibilities (incl. enforcement & coordination)	Cost Estimates	
Construction Phase	Vegetation removal and possible habitat loss; dredging and filling of wetlands or coastal areas	Include in determination of overall environmental protection measures for selection of alternative sites and layout to avoid loss of ecological resources; restore or create similar vegetation, habitats or wetlands/coastal areas	Identification of agencies involved in natural resource and /environmental management and agreement on responsibilities for implementing protection measures; consultants	Costs of mitigation in light of inst. responsibilities to be included in design preparation, including cross-sectoral management	Potential for impacts on fisheries, rare or endangered species.
	Increased stream sediment from erosion at road cuts, construction sites and waste dumps; borrow pits	Protect susceptible surfaces with mulch or fabric, and plant erodible surfaces. Limitation of earthmoving to dry periods. Installation of sedimentation basins, seeding or planting of erodible surfaces; conduct in tandem with installation of culverts, ridges, etc. and flow regulating works	Agencies involved in project implementation; contractors and consultants; environment/water resources agencies	Costs against involved agencies' responsibilities and in construction costs	Any risks posed by neighboring development activities
	Air quality: impacts of dust, machinery, etc	Establish a dust suppression operation and maintain efficient fleet of trucks, etc. Initiate air monitoring program in site vicinity to collect pre-operation baseline, measure plant impacts when it comes on line and to develop a database for future energy projects	Agencies involved in project implementation; contractors and consultants; environment agency.	Costs against involved agencies' responsibilities and in construction costs	
	Increased traffic and disruption to communities	Develop a traffic plan that includes phasing road use by workers; upgrade roads and intersections.	Agencies involved in project implementation; contractors and consultants.	Costs against involved agencies' responsibilities and in construction costs	Any risks posed by neighboring development activities
	Occupational health and safety of workers and protection of communities	Develop a plan for worker protection and for control of HIV/AIDS; involvement of local communities.	Agencies involved in project implementation; contractors and consultants.	Costs against involved agencies' responsibilities and in construction costs	
	Noise and vibration	Implement agreed external noise and vibration recommendations to reduce impacts, eg, control the timing equipment usage to least disruptive periods, install noise barriers	Agencies involved in project implementation; contractors and consultants	Costs against involved agencies' responsibilities and in construction costs	
	Induced secondary development, including demands upon infrastructure	Provide infrastructure plan and financial support to meet increased demands; construct facilities to reduce demand sin context of regional planning.	Regional and local planning authorities; agencies involved in project implementation		

Operation and Maintenance Phase	Atmospheric emissions: potential impacts upon fauna and flora, habitats, agriculture and human health; transboundary pollution and impacts upon global climate.	Locate facility away from sensitive areas; design higher stacks to reduce ground level concentrations; use cleaner fuels, install air pollution control equipment.	Agencies involved in project implementation; contractors and consultants; environment agency	Costs against involved agencies' responsibilities and in construction costs
	Changes to runoff levels	Maintenance of surface drainage and application of standards and regulations	Sector agency and environment/water resources agencies; contractors and consultants	Costs against involved agencies' responsibilities and in construction costs.
	Consumption of water	Reduction in volumes available for drinking, irrigation, and navigation and other uses in water short areas.	Sector agency and environment/water resources agencies; planning authorities; contractors and consultants	
	Water discharge: changes to surface and groundwater quality.	On-site chemical or mechanical treatment; use of liners to prevent groundwater contamination; deep well injection below potable zones; dilution of effluent at point of discharge.	Sector agency and environment/water resources agencies; contractors and consultants	Costs against involved agencies' responsibilities and in construction costs.
	Chemical discharges and spills: toxic impact; disposal of wastes	Develop spill prevention plans; develop traps and containment systems and chemically treat discharges on-site; provide a waste management plan for collection, handling and recycling or proper disposal.	Sector agency and environment/water resources agencies; contractors and consultants	Costs against involved agencies' responsibilities and in construction costs.
	Heated water discharge: impacts on aquatic life	Use alternative heat dissipation design (eg, closed cycle cooling); dilute thermal condition by discharging water into larger receiving water body; install mechanical diffusers; cool water in on-site holding pond(s) prior to discharge; evaluate opportunities to use waste heat.	Sector agency and environment/water resources agencies; contractors and consultants	Costs against involved agencies' responsibilities and in construction costs
	Water intake: entrainment and impingement of aquatic organisms.	Position water intake in area to avoid significant impacts; install screens to eliminate entrainment and impingement	Sector agency and environment/water resources agencies; contractors and consultants	Costs against involved agencies' responsibilities and in construction costs
	Worker exposure to dust from ash and coal; toxic gases leaking from boilers; excessive noise	Provide dust collector equipment and maintain low dust levels; provide masks when necessary; maintain boilers and monitor levels so they not exceed set levels; maintain noise levels below 90 dBA or provide ear protection; occupational health and safety planning and training	Sector agency and environment agency; contractors and consultants	Costs against involved agencies' responsibilities and in construction costs
	Industrial risks and emergency response	Adoption of findings of the Major Hazard Assessment: engineering controls, administrative controls, personnel protection, etc.	Local authorities; sector agency and environment agency; contractors and consultants	Costs against involved agencies' responsibilities and in construction costs

Source: World Bank Guidelines, 2006

Table B–2. Thermal Power Projects Template, Part II—Monitoring Plan

Project Activity	Parameters To be Monitored	Locations	Measurements (incl. methods & equipment)	Frequency Of Measurement/ Reporting	Responsibilities (incl. review and reporting)	Cost (Equipment & individuals)
Pre-Construction Phase (baseline)	Determination of key variables to assess potential impacts on air quality and water quality and quantity and sensitive habitats. Note: Geophysical testing may be required to characterize geological conditions under the proposed facility	Determination of air sheds. Determination of key rivers and streams of impacted watersheds and sensitive habitats.	Determination in relation to atmospheric/ meteorological conditions; determination of hydrological regime monitoring requirements	Field visit with agencies to determine needs in relation to atmospheric and hydrological regime and flows	Identification of agencies involved in air and water resources management and agreement on responsibilities for monitoring protection measures; consultants	Costs of monitoring in light of inst. responsibilities to be incorporated in design preparation, including cross-sectoral management
Construction Phase	Particulate matter	Minimum of 4/6 locations, including sensitive areas: nearby residences, crops		Every 2/4 weeks	Agencies involved in project implementation; contractors and consultants; environment agency.	Costs against involved agencies' responsibilities and in construction costs
	Water quality	At all discharges, including runoff and wastewater.	pH, BOD, COD ,oil and grease, suspended solids, selected heavy metals and total and fecal coliform	Regular check and according to changes to runoff/ incidents	Sector agency and environment/water resources agencies	Costs against involved agencies' responsibilities and in construction costs
	Noise and vibration	A minimum of 6 locations: nearest residences and site boundary	Decibels (dBA)	Regular check	Sector agency and environment agency	Costs against involved agencies' responsibilities and in construction costs
Operation/ Maintenance Phase: Coal and oil plants Gas plants	Air quality: point and no-point emissions Air quality: point and no-point emissions	Stack exhaust and upwind and downwind of emissions Stack exhaust and upwind and downwind of emissions	Particulate matter, sulfur and nitrogen dioxides and heavy metals in flue gases; installation of automatic air quality monitoring systems measuring ambient levels of particulates, sulfur and nitrogen dioxides outside plant boundary or where there are sensitive receptors	Continuous/ 24 hour average Continuous/ 24 hour average	Sector agency and environment agency.	Costs against involved agencies' responsibilities and in maintenance costs

		Nitrogen dioxide in flue gases; installation of automatic air quality monitoring systems measuring ambient levels of nitrogen dioxide outside plant boundary where maximum ambient concentration is expected or where there are sensitive receptors			
Water quality of discharged water. Note: Groundwater monitoring may be required if contamination of groundwater is predicted.	In discharged water and upstream and downstream from point of discharge in any receiving water body used by the public or considered sensitive, eg rivers, drinking and irrigation supplies.	pH, BOD, COD ,oil and grease, total suspended solids, selected heavy metals and total and fecal coliform; total residual chlorine, temperature at edge of mixing zone	Continuous (with alarm in some cases) or weekly depending upon the parameter.	Sector agency and environment agency	Costs against involved agencies' responsibilities and in maintenance costs
Aquatic organisms	Upstream and downstream from point of discharge in any receiving water body	Sampling of representative species of organisms	Biweekly	Sector agency and environment agency	Costs against involved agencies' responsibilities and in maintenance costs
Noise and vibration	6-10 sites at nearest residences around the plant	Decibels (dBA)	Monthly	Sector agency and environment agency	Costs against involved agencies' responsibilities and in maintenance costs
Health and safety	At selected sites in workplace posing risks	Regular and spot dust sampling; monitor ambient levels so toxins do not exceed set levels; dBA levels; check on provision of masks and ear defenders; medical monitoring; adequacy of occupational health and safety training.	Daily/bi-weekly depending upon selected site.	Sector agency and environment agency	Costs against involved agencies' responsibilities and in maintenance costs
Industrial risks and emergency response	At selected sensitive areas	Test preparedness of emergency plans, including evacuation; update profile of risks of fire and explosion; liaison with local authorities.	Daily/bi-weekly depending upon selected site	Local authorities; sector agency and environment agency	Costs against involved agencies' responsibilities and in maintenance costs

Source: World Bank Guidelines, 2006

Glossary

Advance payment guarantee: A financial instrument that assists the contractor in purchasing and assembling the materials, equipment, and personnel necessary to begin construction, to meet the requirements for receipt of progress payments under the contract.

American depository receipt (ADR): A certificate of ownership, issued by a U.S. bank, representing a claim on underlying foreign securities. ADRs may be traded in lieu of the actual underlying shares.

Associated gas: *Natural gas* found dissolved in or together with crude oil in a reservoir.

Average discounted cost: See *average levelized cost.*

Average incremental cost: Measures the average cost of an increment to the generating capacity which is calculated as the average levelized cost of the incremental unit.

Average levelized cost (also called *average discounted cost* or *average incremental cost*): Measures the average capital and fuel costs for each type of generating plant by dividing the present value of the cost stream by the present value of the output stream.

Balance of payments: The measure of all international financial transactions between residents of a given country and foreign residents.

Balance of trade: The measure of the foreign exchange value of exports minus the value of imports for a given country.

Barrel (bl), petroleum: A unit of volume equal to 42 U.S. gallons.

Basis swap: An agreement in which one counterparty agrees to pay a floating price based on a commodity index at a certain delivery point in exchange for the floating price on the commodity futures market.

Bid bond: A bond required of bidders at the outset of a project to ensure that each bidder is serious and would accept the award of the contract if offered.

Bill of lading (B/L): A contract between a common carrier and a shipper to transport goods to a named destination. The B/L is also a receipt for the goods.

Black market: An illegal foreign exchange market.

Build-own-operate (BOO) and build-operate-transfer (BOT) schemes: Methods by which private sector participation in the energy sector is encouraged. Under these approaches, a project company under private ownership or a joint venture with minority public participation is set up to plan, finance under limited recourse, design, construct, and operate energy generation facilities. In a BOT arrangement, ownership of the facility is ultimately transferred to another entity after a specified period of operation. A variant is the build-lease-transfer (BLT) scheme.

Book runner: The managing or lead underwriter who maintains the books of securities sold for a new issue.

Branch: A foreign operation not incorporated in the host country, as opposed to a subsidiary.

British thermal unit (BTU): The quantity of heat needed to raise the temperature of one pound of water by 1°F at or near 39.2°F. (See *heat content of a quantity of fuel, gross,* and *heat content of a quantity of fuel, net.*)

Buyback rate: The rate charged by a utility to a private power developer for buying back some or all of the power that the developer is obligated to deliver.

Capacity factor: The ratio of the electrical energy produced by a generating unit for a given period of time to the electrical energy that could have been produced at continuous full-power operation during the same period.

Capital account: That portion of the balance of payments that measures public and private international lending and investment.

Capital budgeting: Analytical approach used to determine whether investment in long-lived assets or projects is viable.

Capital markets: The financial markets in various countries in which different types of long-term debt and/or ownership of securities—or claims on those securities—are purchased and sold.

City gate: A point or measuring station at which a gas distribution utility receives gas from a natural gas pipeline company or transmission system.

Commercial bank: A term historically used to distinguish traditional bank services from investment banking, referring to the process of collecting funds from depositors and making loans to businesses and consumers.

Commercial risk: A category comprising risks under the control of the owner of the project and risks of project development, timely construction, and efficient and economical operation and maintenance. (See also *force majeure risk, political risk,* and *project risk.*)

Common financing: Financing raised for a project as a whole and backed by several undertakings of the sponsors and appropriate interpartner default clauses. Common financing is often used in the power sector (and is distinct from individual financing, usually used in oil projects, in which each partner is individually responsible for financing).

Concession: An arrangement whereby a private party leases assets for service provision from a public authority for an extended period and has responsibility for financing specified new fixed investments during the period. The assets revert to the public sector at expiration of the contract. (See also *franchise.*)

Concessional financing: See *soft loan.*

Concessional loan: See *soft loan.*

Consolidation: In the context of accounting for multinational corporations, the process of preparing a single reporting-currency financial statement, combining financial statements of affiliates that are in fact measured in different currencies.

Cost and freight (C&F): Exporter's quoted price, including the cost of transportation to the named port of destination.

Cost, insurance, and freight (CIF): Exporter's quoted price, including the cost of packaging, freight or carriage, insurance premium, and other charges, from the time of loading in the country of export to their arrival at the named port of destination or place of transshipment.

Cost-of-service tariff: See *throughput contract.*

Cost risk: Category comprising the risks of cost increases caused by inflation, fluctuation of interest and foreign exchange rates, changes in availability of materials and fuels, cost overruns, and project delays (combines with *revenue risk* into *margin risk*).

Counterguarantee: A third-party guarantee used when one party purchases a service or a product for monetary payments and this party's ability to make payments is in question. This guarantee is required by the provider of the service or product, to ensure that payments will be made.

Creeping expropriation: *Expropriation* as a result of the growth of a *host country*'s take through increased taxes, royalties or fees during project construction or operation.

Cross-currency swap: A transaction in which two counterparties agree to exchange principal and interest denominated in different currencies based on an agreed currency exchange rate.

Cross-subsidy: The allocation of funds provided by one or more products or sectors of the economy to other products or sectors of the economy. Often, this process is not transparent (e.g., high prices for industrial users of electricity can be used to provide subsidies to domestic consumers).

Cubic foot (natural gas): A unit of volume equal to one cubic foot at a pressure base of 14.73 pounds standard per square inch absolute and a temperature base of 60°F.

Current account: In the balance of payments, the net flow of goods, services, and unilateral transfers (e.g., gifts) between a country and all foreign countries.

Debt ratio: See *gearing*.

Debt service: Periodic payment of principal and interest on loans, bonds, or fixed- or floating-rate notes.

Deficiency agreement: See *throughput contract*.

Depletion premium: Used in calculating the economic cost of supply of an exhaustible resource and indicates the opportunity cost of consuming a unit of the resource now, rather than at a given point in the future, taking into account the cost of alternative fuels. In a country rich in a given resource, the depletion premium is negligible; in a country already short of the resource, the point of switching to an alternative fuel becomes the present, and the economic analysis is based on the *net back value*.

Devaluation: A government action to reduce the purchasing power or value of local currency against convertible currencies.

Development well: A well drilled within the proven area of an oil or gas reservoir to the depth of a stratigraphic horizon known to be productive.

Economic analysis: An analysis that takes into account the costs and benefits of a project to society or the country as a whole. For example, if electricity were subsidized, an economic analysis would use as its basis the price paid by the consumer plus the subsidy (cf. *financial analysis*).

Electrical system energy losses: The amount of energy lost during generation, transmission, and distribution of electricity, including plant and unaccounted-for uses.

Electric power plant: A station containing prime movers, electric generators, and auxiliary equipment for converting mechanical, chemical, and/or fission energy into electric energy.

Electricity generation: The process of producing electric energy or transforming other forms of energy into electric energy; also, the amount of electric energy produced, expressed in watt-hours (Wh).

Electricity generation, gross: The total amount of electric energy produced by the generating station or stations, measured at the generator terminals.

Electricity generation, net: Gross generation minus electricity consumed at the generating plant for station use. Electricity required for pumping at pumped-storage plants is regarded as plant use and is deducted from gross generation.

End-use sectors: The residential, commercial, industrial, and transportation sectors of the economy.

Escrow account: A special account, often outside the project's *host country,* to which all or part of project revenues are channeled, to ensure disbursement of funds according to agreed conditions and priorities. When all funds are deposited, the disbursement first covers operating cost and then debt service before remitting the remainder to the project company. An escrow account agreement terminates when all debt obligations have been fully paid.

Euro area: The countries that have adopted the euro. The list of countries in the euro area includes Austria, Belgium, Finland, France, Germany, Greece, Ireland, Italy, Luxembourg, Netherlands, Portugal, and Spain and will expand as more countries adopt the euro as their currency.

Eurobond: A bond sold outside the country in whose currency it is denominated.

Eurocurrency: A currency deposited in a bank located in a country other than the country issuing the currency.

Eurodollar: A U.S. dollar deposited in a bank outside the United States.

European currency unit (ECU): Composite currency created by the European Monetary System to function as a reserve currency numeraire for denominating a number of financial instruments.

Exchange rate: The price of a unit of one country's currency expressed in terms of the currency of another country.

Export parity: Price minus associated transportation cost.

Expropriation: Forced transfer of ownership from a private owner to a government institution. (See also *creeping expropriation.*)

Financial analysis: An analysis that takes into account the costs and benefits of a project to the entity undertaking it. For example, if electricity is subsidized, the financial analysis would consider only the revenue to the entity and not the overall benefit of any subsidy given to the buyer (cf. economic analysis).

Financial closing: Occurs when all the conditions of lenders and investors have been met and financing disbursements can take place.

Financial closure: The point at which the legal and financial commitments for a project are formally executed, generally followed soon after by disbursement of funds and the commencement of project construction.

Financial derivatives: Instruments such as *interest-rate swaps* and caps, as well as *currency swaps* and options used to hedge against unprotected increases in financial costs—in particular, those caused by an increase in interest rates or an adverse movement in the exchange rate.

Financial engineering: The complex discipline of combining various instruments for securing guarantees, borrowing, and mobilizing equity toward the end of financing a project.

Fixed exchange rate: Foreign exchange rate set and maintained by government support.

Floating exchange rate: Foreign exchange rate determined by demand and supply in an open market presumably free of government interference.

Force majeure risk: Danger from unavoidable events, such as tornadoes, earthquakes, hurricanes, and lightning, that (combined with *political risk* and *commercial risk*) constitutes part of total *project risk.*

Forward foreign exchange contract: A contract to pay or receive specific amounts of a currency at a future date in exchange for another currency at an agreed exchange rate.

Forward rate agreement: An agreement to exchange dollar amounts at a specified future date on the basis of the difference between a particular interest-rate index and an agreed fixed interest rate.

Fossil fuel: Any naturally occurring organic fuel, such as *petroleum,* coal, and *natural gas.*

Franchise (also called *concession*): The grant of certain rights to an individual group, partnership, or corporation.

Free on board (FOB): International trade term in which the exporter's quoted price includes the cost of loading goods into transport vessels at a named point.

Futures contract: A standardized contract, traded on an organized exchange, that obligates one party to buy and another party to sell a specific asset at a future date.

Gulf Cooperation Council (GCC): The GCC was established in 1991 with the purpose of increasing integration between the member countries. The GCC has six members: Bahrain, Kuwait, Oman, Qatar, Saudi Arabia, and the United Arab Emirates. Yemen is being considered for future membership.

Gearing (also called *debt ratio* or *leverage*): The ratio of a project's fixed-interest debt to its equity plus debt. A highly geared (or leveraged) project is one in which the proportion of debt is comparatively great. Energy projects are typically financed in the range of 20%–40% equity and 60%–80% debt. A project with a 30% equity component thus would have a gearing of 70%. Lenders prefer lower gearing, which means greater contributions of equity—and, hence, greater commitment by project sponsors and stockholders—whereas project sponsors prefer higher gearing, to minimize the capital that they must lock into the project.

Gross domestic product (GDP): The total value of goods and services produced within a country over a given period, usually a year; also, the total value of goods and services produced by labor and property located in a country.

Gross national product (GNP): Comprises *GDP* plus income to domestic entities from foreign holdings minus income to foreign entities from domestic holdings.

Gross product worth (GPW): The weighted average value of the refinery yield—that is, the mix of refined products that are derivable from a given quantify of crude. GPW is obtained by multiplying the prevailing price of the derived products by their percentage yield from the crude product (cf. *net product worth* and *net back value*).

Hard currency: All major convertible currencies, such as the U.S. dollar, the British pound, the German mark, the Japanese yen, the French franc, the Swiss franc, the Italian lira, and the Dutch guilder.

Heat content of a quantity of fuel, gross: The total amount of heat released when a fuel is burned. Coal, crude oil, and *natural gas* all include chemical compounds of carbon and hydrogen. When those *hydrocarbon* fuels are burned, carbon and hydrogen combine with oxygen in the air to produce carbon dioxide and water. Some of the energy released in burning goes into transforming the water into steam and is usually lost. The amount of heat spent in transforming the water into steam is counted as part of gross heat content but is not counted as part of net heat content.

Heat content of a quantity of fuel, net (also called *lower heating value*): The amount of usable heat energy released when a fuel is burned under conditions similar to those in which it is normally used.

Heavy oil: The fuel oils remaining after the lighter oils have been distilled off during the refining process. Except for startup and flame stabilization, virtually all *petroleum* used in steam electric power plants is heavy oil.

Home country: The country in which a *private power developer* is registered.

Host country: The country in which an energy project is undertaken.

Hydrocarbon: An organic chemical compound of hydrogen and carbon in the gaseous, liquid, or solid phase. The molecular structure of hydrocarbon compounds varies from the simplest (methane, the primary constituent of *natural gas*) to the very heavy and very complex, e.g., crude oil of various quality.

Hydroelectric power: Production of electricity from the kinetic energy of falling water.

Ijara: Bonds that are backed by an asset under a leasing contract.

Implementation agreements: Project-specific agreements, required for successful project development and allocation of risk, that provide government assurances and guarantees to private energy producers.

Implementation team: The *private power developer* and its contractors.

Independent power producer (IPP): A private energy producer that has developed power plants, typically on a project-finance basis, to sell power to an existing utility or directly to distributors or large consumers.

Indexed tariff: An adjusted tariff, based on a variable such as periodic fuel price, interest rates (local or foreign), exchange rate, or inflation rate.

Initial public offering (IPO): The first sale of shares of an enterprise to the public.

Interest-rate swap: An agreement in which one counterparty agrees to pay a fixed rate of interest to the other counterparty in exchange for a variable rate of interest on a fixed notional principal amount over a specified period of time.

Internal combustion electric power plant: A power plant in which the prime mover is an internal combustion engine. Diesel or gas-fired engines are the principal types used in electric power plants, which are usually operated during periods of high demand for electricity.

Internal rate of return (IRR): The discount rate that equates the present value of a project's expected cash inflows to the present value of the project's expected costs.

International oil company (IOC): Category encompassing integrated major oil companies, large independents, and large exploration and production companies. (Cf. *national oil companies*.)

Investment bank: See *merchant bank*.

Investors: Individuals, groups, or companies that invest cash in a private energy developer, group, or company.

Irrevocable letter of credit: A guarantee by a commercial bank to issue credit on demand, up to a specified maximum limit. This service is provided by banks for a fee.

Istisna'a: A manufacturing contract compliant with Islamic law.

Joint venture: A business venture that is owned by two or more other businesses. Often, the business owners are from different countries.

Kabushiki-kaishi (KK): Japanese term for stock company.

Lead bank: A bank that arranges the loan, takes part of the loan, and syndicates the rest of the loan to other banks.

Letter of credit (L/C): An instrument issued by a bank, in which the bank promises to pay a beneficiary on presentation of documents specified in the L/C.

Leverage: The amount of borrowed capital used to finance assets. (See also *gearing*.)

Limited-recourse financing: A lending arrangement under which repayment of a loan and recourse in the event of a default relies mainly on the project's cash flow. Project sponsors remain responsible for covering a default during a limited period of time.

Liquefied natural gas (LNG): *Natural gas* (primarily methane) that has been liquefied by reducing its temperature to 260°F at atmospheric pressure.

Liquefied petroleum gases (LPG): Ethane, ethylene, propane, propylene, normal butane, butylene, and isobutane produced at refineries or natural gas processing plants, including plants that fractionate new *natural gas plant liquids.*

London Interbank Offer Rate (LIBOR): The deposit rate applicable to interbank loans within London.

Lower heating value: See *heat content of a quantity of fuel, net.*

Lump-sum turnkey (LSTK) contract: A contract that ensures completion of a project on time, within budget, and within acceptable standards of operation. LSTK contracts involve a set sum for completion of a project but have stringent reward and penalty clauses that protect sponsors against cost overruns and operational deficiencies. (See also *turnkey contract.*)

Maintenance bond: A source of funds for correcting construction or performance defects discovered after completion of project construction. Typically, the *performance bond* and the *retention bond* are converted to maintenance bonds on completion of the contract.

Margin risk: The sum of *cost risks* and *revenue risks.*

Merchant bank (also called *investment bank*): A bank that deals mostly with international finance, lending to corporations, and underwriting and does not provide regular banking services to general public.

Mezzanine financing: Subordinated loans and issues that fall between debt and equity. In certain cases, mezzanine financing gives lenders the right to take an equity stake if interest is not paid on time.

Mixed credits: A blend of credits from different sources and at different levels of *concessional finance*.

Murabaha: Bonds compliant with Islamic law that are based on a cost-plus sale contract.

Musharaka: A profit-and-loss–sharing contract compliant with Islamic law.

Naphtha: A generic term applied to a petroleum fraction with a boiling point in the range of 122–400°F.

National oil company (NOC): A category that encompasses state-owned oil companies.

Natural gas: A mixture of hydrocarbons (principally methane) and small quantities of various nonhydrocarbons existing in the gaseous phase or in solution with crude oil in underground reservoirs.

Natural gas, dry: The marketable portion of natural gas production, which is obtained by subtracting extraction losses, including natural gas liquids removed at natural gas processing plants, from total production.

Natural gas plant liquids (NGPL): Natural gas liquids recovered from natural gas in processing plants and, in some situations, from natural gas field facilities, as well as those extracted by fractionators.

Natural gas, wet: Natural gas prior to the extraction of liquids and other miscellaneous products.

Net back value: An estimate of the value of products that a given quantity of a specific type of crude oil or *natural gas* will yield after passing through a particular configuration of *refinery* or processing plant. Net back value is obtained by subtracting the *net product worth* from the *gross product worth*. Further subtraction of the costs of transportation and insurance gives the net back value at the port of loading.

Net present value (NPV): An approach to capital budgeting in which the present value of expected future case inflows is subtracted from the present value of outflows.

Net product worth (NPW): Derived by subtracting from the *gross product worth* of a given quantity of oil or gas the operating expenses involved in handling the last (marginal) quantity of the crude product (not including amortization and depreciation).

Nonrecourse financing: Recourse, belonging exclusively to the project company, for debt repayment, default, or both.

Nuclear electric power: Electricity generated by an electric power plant whose turbines are driven by steam generated by heat from the fissioning of nuclear fuel in a reactor.

OECD Consensus: Guidelines intended to prevent distortions in price competition among manufacturers of different countries. The OECD Consensus is derived from the *Organisation for Economic Co-operation and Development*'s (OECD) "Arrangement on Guidelines for Officially Supported Credits" (1978), which limits export credit to 85% of the contract value and holds interest rates to a minimum of the OECD interest-rate matrix, which is revised semiannually.

Oil well: A well completed for the production of crude oil from one or more oil zones or reservoirs. Wells producing both crude oil and *natural gas* are classified as oil wells.

Onlending (also called *relending*): Lending through financial intermediaries.

Option: A contract that provides the option holder the right, but not the obligation, to buy or sell the underlying instrument.

Organisation for Economic Co-operation and Development (OECD): A group of countries that have set up agencies that provide financial and technical assistance to developing countries in the form of grants and *concessional loans.* Current members are Australia, Austria, Belgium, Canada, Denmark, Finland, France, Germany, Greece, Iceland, Ireland, Italy, Japan, Luxembourg, the Netherlands, New Zealand, Norway, Portugal, Spain, Sweden, Switzerland, Turkey, the United Kingdom, and the United States and its territories (Guam, Puerto Rico, and the U.S. Virgin Islands).

Organization of Petroleum Exporting Countries (OPEC): A group of countries that have organized for the purpose of negotiating with oil companies on matters of oil production, prices, and future concession rights. Current members are Algeria, Gabon, Indonesia, Iran, Iraq, Kuwait, Libya, Nigeria, Qatar, Saudi Arabia, the United Arab Emirates, and Venezuela.

Over the counter (OTC) market: Trading of securities through a network of traders, rather than on a formal exchange.

Parallel financing: Arrangement of separate loan agreements for each type of financing in support of a project.

Performance bond: Additional funds in the event a contractor fails to perform for any reason. The existence of such a bond is also an endorsement of the credit and confidence of the guarantor in the ability and professional standing of the contractor. Performance bonds are purchased by the project developer and issued by *commercial banks* or insurance companies to guarantee full and successful implementation of a contract according to prespecified performance guidelines.

Petrochemical feedstocks: Chemical feedstocks derived from *petroleum* principally for the manufacture of chemicals, synthetic rubber, and a variety of plastics.

Petroleum: Oil and oil products in all forms, such as crude oil, lease condensate, unfinished oils, *petroleum products, natural gas products liquids,* and nonhydrocarbon compounds blended into finished petroleum products.

Petroleum products: Products obtained from the processing of crude oil (including lease condensate), natural gas, and other *hydrocarbon* compounds. Petroleum products include unfinished oil, *liquefied petroleum gases,* pentanes plus, aviation gasoline, motor gasoline, *naphtha*-type jet fuel, kerosene-type jet fuel, kerosene, distillate fuel oil, residual fuel oil, petrochemical feedstocks, special naphthas, lubricants, waxes, petroleum coke, asphalt, road oil, still gas, and miscellaneous products.

Pink sheet stocks: Stocks traded *over the counter*. Companies that are quoted on the pink sheet system do not need to file with the U.S. Securities and Exchange Commission.

Pipeline fuel: Gas consumed in the operation of pipelines, primarily in compressors.

Plant downtime: Time when a plant is not producing power because of scheduled or forced outage or shutdown.

Political risk: Risks stemming from political uncertainty or difficulty in the *host country,* including *expropriation* of project property, breach of undertakings by the host government or government entities, civil unrest or revolution, war, prevention of profit repatriation, and inconvertibility of local currency.

Power purchaser: The entity purchasing power from a *private power developer.* Usually, the public utility of the *host country* is the power purchaser.

Price swap: An agreement in which one counterparty agrees to pay a fixed price for a commodity to the other counterparty in exchange for a price based on a commodity index. The payments are based on fixed notional quantities.

Private power developer: An individual group or company that develops power plants on a private basis to own, operate, lease, or transfer.

Project company: The special-purpose entity that assumes legal and financial responsibility for construction and operation of the project. Recourse is limited to the project company.

Project risk: The total risk, including *commercial risk, political risk,* and *force majeure risk.*

Put option: Allows holders to present—or "put"—bonds for repayment at maturity.

Put-or-pay contract (also called *supply-or-pay contract*): Provided by suppliers of energy, raw materials, or products to projects needing assured inputs over long periods, at predictable prices, to meet production cost targets. The put-or-pay obligor must either supply the energy, raw material, or product or pay the project company the difference in cost incurred in obtaining the input from another source.

Qualified institutional buyers (QIBs): Mainly institutions that manage over $100 million in assets. These investors are considered to have the ability to assess the risk of investing.

Quasi-equity: A special type of funding meant to attract risk-averse investors to a project. It is issued in the form of preferred shares, whose holders are paid before holders of ordinary shares but after creditors and holders of debt financing.

Refinery, petroleum: An installation that manufactures finished *petroleum products* from crude oil, unfinished oils, natural gas liquids, other *hydrocarbons,* and alcohol.

Relending: See *onlending.*

Retention money bond: A bond that contractors substitute for the portion of the progress payment to a contractor that sponsors would otherwise retain to cover unforeseen expenses caused by any contractor mistakes in construction.

Revenue risk: Risk to a project from fluctuations in demand and price (combines with *cost risk* into *margin risk*).

Riba: An Arabic word meaning interest or usury. Riba is prohibited in Islamic law.

Risk profile: The level of risk attributable to political, economic, or financial uncertainty to which an investor is exposed. This determines the rate of return that an investor requires in order to tolerate exposure to the level of adversity in any country or project.

Rule 144A: Allows the sale of restricted investment-grade securities (notes or bonds) to *qualified institutional buyers* without registration with the U.S. Securities and Exchange Commission, thereby providing foreign companies with access to the U.S. capital market. The issue of these securities does not require the same detailed financial information as a public offering. However, three years after their issue, 144A securities can be freely traded in the U.S. market.

Samurai bond: A yen-denominated bond issued in Japan by a non-Japanese company.

Senior loan: A loan having priority in repayment over other loans or obligations.

Sharia: The body of Islamic law.

Soft financing: See *soft loan.*

Soft loan (also called, *concessional financing, concessional loan,* and *soft financing*): A loan or financing arrangement bearing no interest or interest below the cost of the capital lent. Loans by the World Bank are somewhat below commercial rates but are not soft. The International Development Association (IDA), a World Bank Group affiliate, lends to the world's poorest countries on soft, or concessional, terms.

Sovereign guarantee: Government guarantee (e.g., guarantee of repayment of a loan by a state entity or of the obligations of a purchasing utility under a power purchase agreement).

Sovereign rating: Measure of the ability and willingness of a country to service its debts. Credit-rating agencies may interpret a country's sovereign rating as a ceiling for the rating of any other entity domiciled in the country, recognizing that the central government's control over economic, fiscal, monetary, and trade policy renders the country's credit standing higher than that of any other debtor in the country.

Sponsors' completion guarantee: A guarantee that commits the sponsors to completing the project within a certain time period and to providing the funds to pay all cost overruns. However, other guarantees (e.g., *performance bonds* and *maintenance bonds* by contractors and equipment suppliers) are normally sought to protect the project sponsors and to provide financiers with additional assurances that project completion risks are managed.

Substitute natural gas: See *synthetic natural gas.*

Sukuk: A *Sharia*-compliant capital market product.

Supply-or-pay contract: See *put-or-pay contract.*

Syndicated bank facility: See *syndicated loan.*

Syndicated loan (or *syndicated bank facility*): A very large loan in which a group of banks work together to provide funds for one borrower.

Synthetic natural gas (SNG; also called *substitute natural gas*): A manufactured product chemically similar to natural gas, resulting from the conversion or re-forming of petroleum *hydrocarbons.* SNG may easily be substituted for, or interchanged with, pipeline-quality natural gas.

Take-or-pay contract: An unconditional contractual arrangement, between the project sponsor and project's customer, that obliges the customer to make periodic payments in the future in return for fixed amounts or quantities of products, goods, or services at specified prices, regardless whether the service or the product is used by the customer. Payments are usually subject to escalation because of inflation. The obligor can protect its interest by retaining rights to take over the project in the event of failure by the supplier to perform, but in doing so, the obligor would also have to assume or pay the debt used to finance the project.

Tax holidays: Exemptions from some or all taxes for a specified period.

Throughput contract (also called *tolling agreement*, *cost-of-service tariff*, or *deficiency agreement*): The equivalent to a *take-or-pay contract* in projects that provide services such as power transmission, power distribution, oil pipeline transportation, or refining—in which the obligor pays regardless whether the service is used. Because the guarantee is unconditional and lasts for the life of the loan, lenders regard throughput contracts as a guaranteed source of income.

Tolling agreement: See *throughput contract.*

Tranche: One of a number of classes of bonds that are offered as part of the same deal.

Turnkey contract: A contract given by the project developer to a prime contractor who will be responsible for design and implementation of a project from start to finish and who will provide a completed, operational project on a stipulated date, on a lump-sum turnkey basis or at cost plus a fixed fee.

Underwriting insurance: The process of measuring risk exposure and determining the premium that needs to be charged to insure a risk.

Underwriting securities: The process by which investment bankers raise investment capital from investors on behalf of corporations that are providing (debt or equity) issues. A syndicate of banks underwrites the transaction. That is, they take the risk of finding purchasers (investors) for the security issue. Should they not be able to find enough investors, they end up holding some securities themselves.

Vented natural gas: Gas released into the air on the base site or at processing plants.

Wellhead price: The value of crude oil or *natural gas* at the mouth of the well.

Wholesale banking: Banking services between *merchant banks* and other financial institutions—in other words, business-to-business banking.

Yankee bond: A U.S. dollar–denominated bond issued in the United States by a non-U.S. company.

Bibliography

The 2005 Natural Gas Year in Review: Cedigaz' First Estimates. 2006. Paris: Cedigaz.

Alami, Randa. 2005. *Financial Aspects of Arab Power Development*. London: Oxford Institute for Energy Studies.

Baughman, David, and Matthew Buresch. 1994. Mobilizing private capital for the power sector: Experience in Asia and Latin America. Joint World Bank–USAID Discussion Paper. Washington, DC: World Bank Cofinancing and Financial Advisory Services Department, World Bank Industry and Energy Department, and USAID Office of Energy, Environment and Technology.

Besant-Jones, John, ed. 1990. Private sector participation in power through BOOT schemes. IEN Energy Series Paper 33. Washington, DC: World Bank, Industry and Energy Department.

Bond, Gary, and Laurence Carter. 1994. Financing private infrastructure projects: emerging trends from IFC's experience. IFC Discussion Paper 23. Washington, DC: International Finance Corporation.

———. 1995. Financing energy projects: Experience of the International Finance Corporation. *Energy Policy.* 23 (December). 967–975.

Borenztein, E., et al. 2006. *Living with Debt.* Washington, DC: Inter-American Development Bank.

BP Statistical Review of World Energy. 2006. London: BP.

Bulletin of Cedigaz Members: The Gas Industry in the Year 2020. 2006. Paris: Cedigaz.

Clifford, Chance. 1991. *Project Finance.* London: IFR Books.

Cornot-Gandolphe, Sylvie, et al. 2003. The challenge of further cost reductions for new supply options (pipeline, LNG, GTL). Paper presented at the 22nd World Gas Conference, Tokyo.

Crawford, J. P. S. Infrastructure project finance in developing countries." In *Project Finance Yearbook (1991/92),* 84–94. London: Euromoney Publications.

Doing Business. Annual. Washington, DC: World Bank.

Dryland, Hugo. 1991. Natural resource finance in the LDC context. In *Project Finance Yearbook (1991/92),* 107–111. London: Euromoney Publications.

Dunkerley, Joy. 1995. Financing the energy sector in developing countries: An overview. *Energy Policy*. 23 (December). 929–939.

EIA. 2003. *The Global Liquefied Natural Gas Market: Status & Outlook*. Washington, DC: U.S. Government Printing Office.

———. 2007. *Country Analysis Briefs*. Washington, DC: U.S. Government Printing Office.

———. 2007. *International Energy Outlook*. Washington, DC: U.S. Government Printing Office.

Eiteman, David, and Arthur Stonehill. 1989. *Multinational Business Finance*. New York: Addison-Wesley.

Energy Balances and Electricity Profiles. Annual. New York: United Nations.

Energy Statistics Yearbook. Annual. New York: United Nations.

Euromoney Institutional Investor. 2006. Tangguh LNG: Slow burn. *Project Finance Magazine*. September. P. 29.

Fabizzi, Frank, and Dessa Fabizzi. 1995. *The Handbook of Fixed Income Securities*. New York: Irwin Publishing.

Fairclough, A. J. 1994. *World Development: Aid and Joint Venture Finance 1994–1995*. London: Kensington Publications.

Fowres, D. 2000. Leasing in project finance. *Journal of Project Finance*. 6 (1). 21–32.

Fraser, Julia. 2005. *Lessons from Independent Private Power Experiences in Pakistan*. Washington, DC: World Bank.

Galitz, I. 1995. *Financial Engineering: Tools and Techniques to Manage Financial Risks*. London: Pitman.

Global Development Finance. Annual. Washington, DC: World Bank.

Global Development Finance: External Finance for Developing Countries. Annual. Vol. I: Analysis and Summary Tables; and Vol. II: Country Tables. Washington, DC: World Bank.

Global Economic Prospects. Annual. Washington, DC: World Bank.

Goldstein, Morris, and Landau Folkers. 1993. *International Capital Markets.* Parts I and II. Washington, DC: International Monetary Fund.

Griffith-Jones, Stephany, and Ana Teresa Fuzzo. 2004. *Alternative Loan Guarantee Mechanisms and Project Finance in Developing Countries*. Sussex: Institute of Development Studies.

Haghighi, Sanam. 2005. Dual pricing of NGLs in Saudi Arabia and the rules of the World Trade Organization on subsidies. *Middle East Economic Survey.* 48 (17). 22–26.

Håmsø, Bjørn, Afsaneh Mashayekhi, and Hossein Razavi. 1994. The international gas trade: Potential major projects. *Annual Review of Energy and Environment*. 19: 37–73.

Harvey, Charles. 1983. *Analysis of Project Finance in Developing Countries*. London: Heinemann.

Himberg, Harvey. 1995. OPIC support to power projects in Asia. *Energy Policy*. 23. 977–980.

Hollis, Sheila Slocum, and John W. Berresford. 1994. Structuring legal relationships in oil and gas exploration and development in frontier countries. In *International Oil & Gas Investment: Moving Eastward?* edited by Thomas W. Wälde and George K. Ndi, 29–59. London: Graham & Trotman.

Humphries, Michael. 1995. The competitive environment for oil and gas financing. *Energy Policy*. 23 (December). 991–1000.

International Energy Agency. 1994. *Electricity Supply Industry: Structure, Ownership and Regulation in OECD Countries*. Paris: International Energy Agency, Diversification Division.

———. 2005. *Projected Costs of Generating Electricity*. Paris: International Energy Agency.

International Petroleum Encyclopedia. Annual. Tulsa: PennWell.

Jechoutek, Karl G., and Ranjit Lamech. 1995. New directions in electric power financing. *Energy Policy*. 23 (December). 941–953.

Jensen, J. J. 2004. *The Development of a Global LNG Market*. Oxford: Oxford Institute for Energy Studies.

Julius, DeAnne S., and Afsaneh Mashayekhi. 1990. *Economics of Natural Gas: Pricing, Planning and Policy*. Oxford: Oxford University Press.

Kayaloff, Isabelle. 1988. *Export and Project Finance*. London: Euromoney Publications.

Konoplyanik, A. 1994. The Russian oil industry and foreign investments: Legal aspects and the problem of business risk." In *International Oil & Gas Investment: Moving Eastward?* edited by Thomas W. Wälde and George K. Ndi, 181–194. London: Graham & Trotman.

Liquefied Natural Gas. 2005. London:Simmons & Co. International, Integrated Oil Research.

Mabro, R. 2006. Egypt's oil and gas: Some crucial issues. Distinguished Lecture Series 25,Cairo: Egyptian Center for Economic Studies.

McKechnie, Gordon. 1983. *Energy Finance*. London: Euromoney Publications.

Merna, Tony, and Cyrus Njiru. 2002. *Financing Infrastructure Projects*. London: Thomas Telford.

Nevitt, Peter K. 1989. *Project Financing*. London: Euromoney Publications.

Onorato, William T. 1995. Legislative frameworks used to foster petroleum development. World Bank Policy Research Working Paper 1420. Washington, DC: World Bank.

Parker, Nicholas. 1993. *Investing in Emerging Economies: A Business Guide to Official Assistance and Finance*. London: Economist Intelligence Unit.

Perkins, Patricio. 1994. Financing of gas projects in developing countries under a privatization scheme: The case in Argentina. Paper presented at the 19th World Gas Conference, Milan.

Pittore, Francesco. 1991. The export credit agency approach to project finance. In *Project Finance Yearbook (1991/92)*, 49–52. London: Euromoney Publications.

Platt, Gordon. 2006. Best banks in project finance. *Global Finance*. December. Vol. 20. 22–27.

Rajas-Suarez, L., and C. Elias. 2006. *From Growth to Prosperity*. Washington, DC: Inter-American Development Bank.

Razavi, Hossein. 1995. Oil and gas financing by the World Bank. *Energy Policy*. 23 (December). 1001–1007.

Razavi, Hossein, and Fereidun Fesharaki. 1991. *Fundamentals of Petroleum Trading*. Westport, CT: Praeger.

Ryrie, Sir William. 1991. Investing in development. In *Project Finance Yearbook (1991/92)*, 27–34. London: Euromoney Publications.

Schur, Michael, and Stephan von Klaudy. 2006. The role of developing country firms in infrastructure. *Public-Private Infrastructure Advisory Facility: Gridlines*. (April): 1–4..

Smith, Arthur L. 1994. Renaissance of the E&P independents. *Petroleum Economist*. (June): 80–84.

Stern, Jonathan. 2006. *The New Security Environment for European Gas*. Oxford: Oxford Institute for Energy Studies.

Thomas Financial Limited. 2003. Umm al-Nar raising the bar for the Middle East. *Project Finance International*. September. 49–55.

Thuman, Albert, and Eric Woodruff. 2005. *Handbook of Financing Energy Projects.* Lilburn: Fairmont Press.

Tordo, Silvana. 2007. *Legal Arrangements and Fiscal Systems for Hydrocarbons*. Washington, DC: World Bank.

UNIDO. 1996. *The Guidelines for Infrastructure Development through Build-Operate-Transfer Projects*. Vienna: United Nations Industrial Development Organization.

United Nations. 2003. *World Investment Report*. UN Conference on Trade and Development, New York.

U.S. Agency for International Development. 1996. *Strategies for Financing Energy Efficiency*. Washington, DC: U.S. Agency for International Development.

Van Meurs, Pedro. 1988. Financial and fiscal arrangements for petroleum development: An economic analysis. In *Petroleum Investment Policies in Developing Countries*, edited by Thomas Wälde and Nicky Beredjick, 47–79. London: Graham & Trotman.

World Bank Group. 2006. *Private Participation in Infrastructure Database*. Washington, DC: International Finance Corporation.

World Development Indicators. Annual. Washington, DC: World Bank.

World Bank and USAID. 1994. Submission and evaluation of proposals for private power generation projects in developing countries. Occasional Paper 2. Washington, DC: World Bank, Industry and Energy Department.

Index

A glossary of terms starts on page 387.

A

C

D

E

F

G

H

J

K

L

M

N

O

P

Q

R

S

T

V

W–X

Y–Z